AF592999

Field Handling of Natural Gas

Fourth Edition

Edited by Jodie Leecraft

Published by
Petroleum Extension Service
Division of Continuing Education
The University of Texas at Austin
Austin, Texas

1987

© 1987 by The University of Texas at Austin
All rights reserved
Published 1954. Fourth Edition 1987
Second Impression 1990
Printed in the United States of America

This book or parts thereof may not be reproduced in any form without permission of Petroleum Extension Service, The University of Texas at Austin.

Brand names, company names, trademarks, or other identifying symbols appearing in illustrations or text are used for educational purposes only and do not constitute an endorsement by the author or publisher.

Catalog No. 3.10040
ISBN 0-88698-127-1

Contents

Foreword

Since 1944 Petroleum Extension Service (PETEX) has developed and conducted a large number of industry-sponsored short courses. In some cases, the written materials used in the short courses have served as the basis of technical manuals published by PETEX, which are designed to be read and studied independently of the short course. *Field Handling of Natural Gas* is one of those manuals. First published in 1954, it followed the initial term of the API-sponsored short course of the same name.

This edition of *Field Handling of Natural Gas* is a rewritten and expanded version of the first edition and more closely reflects the curriculum presented at the School of Gas Technology, a 90-hour course jointly sponsored by American Petroleum Institute and Gas Processors Association. The manual is a valuable instructional guide to the various pieces of equipment used in field handling of natural gas. It not only explains the roles of the equipment but also tells how to maintain it and make it perform at peak efficiency.

It is hard to believe that natural gas was once regarded as a troublesome by-product of crude oil production when one considers how dependent upon it we are today. Indeed, natural gas has come of age and everyone associated with it should know how to handle it safely and efficiently.

Ron Baker
Director, PETEX

Acknowledgments

Information for the third edition of *Field Handling of Natural Gas* was collected by a special group headed by H. J. Haas, chairman of the Advisory Committee for the School of Gas Technology. The school is operated by Petroleum Extension Service and is jointly sponsored by the API Production Department and Gas Processors Association. This new manual, the fourth edition, has been updated where necessary to conform to present procedures, practices, and equipment. In addition, the eighth chapter, "Instruments and Controls," has been completely rewritten, and a glossary has been added.

Bruce Whalen, a retired staff member of PETEX, rewrote the chapter on instruments and controls and is due our sincere gratitude.

Oscar Barron of Exxon Production Research Company reviewed the manuscript in its entirety and offered helpful suggestions for improvement. He has many times offered his time, expertise, and support to PETEX, and we are deeply appreciative.

George H. Donald of Ingersoll-Rand, TurboMachinery Division, responded most helpfully and effectively to our request for rewriting the section on centrifugal compressors and has our sincere thanks.

The following associations and companies were most generous with their help and are acknowledged with gratitude:

Gas Processors Suppliers Association for permission to reprint material from their *Engineering Data Book*

American Gas Association for permission to reprint material from AGA Report No. 3

Black, Sivalls & Bryson for photographs

Ingersoll Rand for photo and illustration

Kimray for photographs

National Supply for photograph

Sivalls, Inc. for photograph

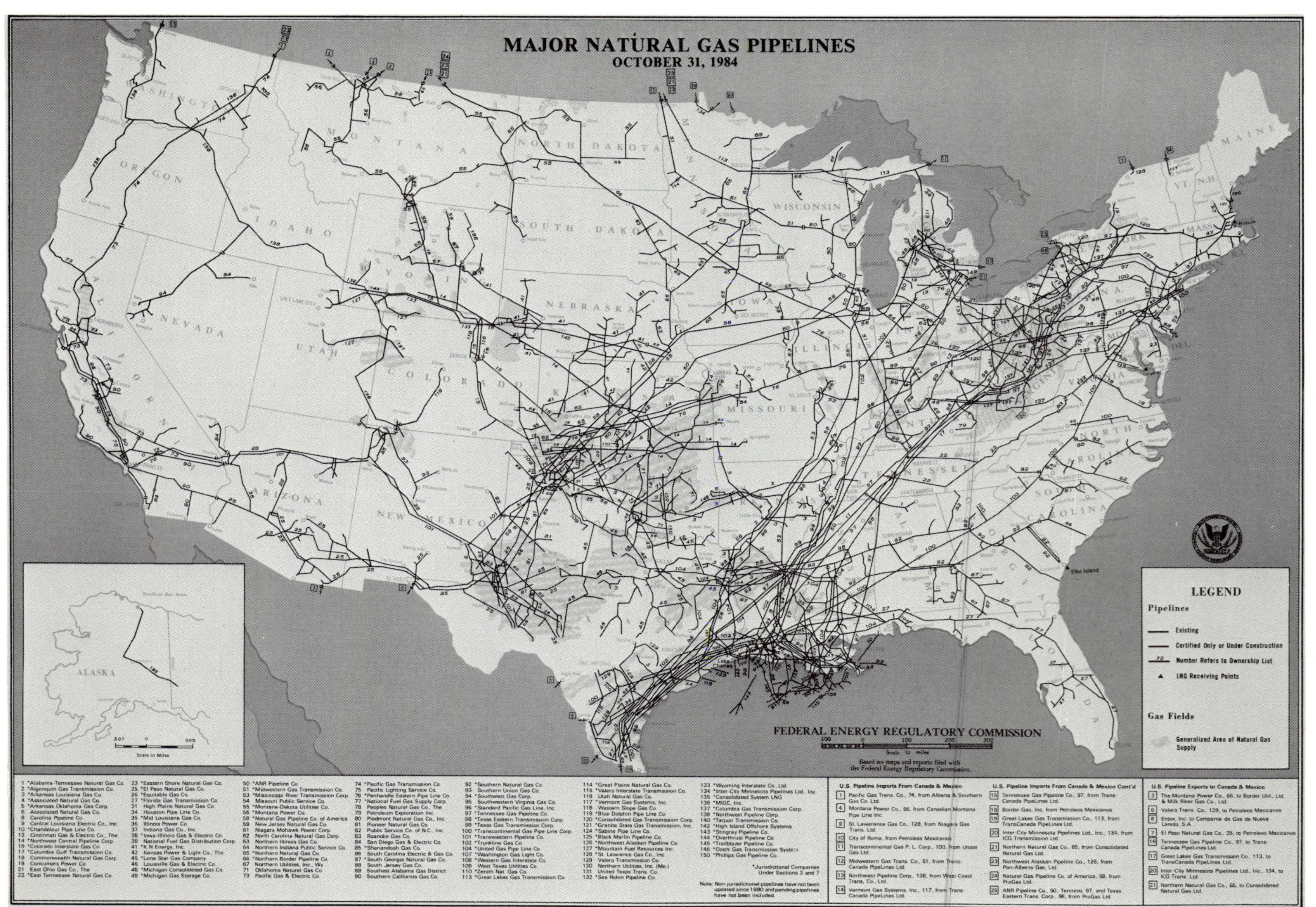

Figure I.1. Major natural gas pipelines in the United States, 1984

Introduction

Natural gas has been used commercially as a fuel for 150 years in America and for centuries in China. The production, processing, and distribution of natural gas has become an important segment of our domestic economy and is a major factor in the world's energy markets.

Since its discovery in the United States, reported to be at Fredonia, New York, in 1821, natural gas has been used as a fuel in areas immediately surrounding the gas fields. In the 1920s and 1930s, a few long pipelines from 22 to 24 inches in diameter, operating at 400 to 600 pounds per square inch, were installed to transport gas to industrial areas remote from the field sources. But during that period also, huge quantities of natural gas that were produced along with crude oil were flared because of the lack of available markets and effective conservation practices.

The modern natural gas industry really began after World War II, when a number of long-distance pipelines were constructed to serve markets in the populated areas of the country. By that time, advances in welding and manufacture of pipe permitted pressures up to 1,000 psi and diameters up to 30 inches. Today virtually every area of the United States is served by natural gas (fig. I.1).

The facilities normally operated in the handling of gas in the field are those required to condition the gas to make it marketable—to remove impurities, water, and excess hydrocarbon liquids, and to control delivery pressure. The last operation involves the use of pressure-reducing regulators or, more often, compressors to raise the pressure.

Natural gas processing plants are usually designed to remove certain valuable products over and above those needed to make the gas marketable—that is, natural gasoline, butane, propane, ethane, and even methane in some instances. Plants may also be designed to recover elemental sulfur from the hydrogen sulfide gas removed from wellhead gas. Another function of plants is to separate the recovered liquid hydrocarbons into various mixtures or pure products by the use of fractionating columns.

Plants nearly always incorporate in their processes many of the functions ordinarily performed by field facilities. Dehydration and H_2S removal form part of the plant operation, and the equipment used is essentially that described in chapters 4, 5, and 6.

1 Characteristics of natural gas

Natural gas is a mixture of hydrocarbon gases, together with some impurities that are the result of decomposed organic material. The impurities also include water vapor and heavier hydrocarbons. When raw natural gas is withdrawn from underground reservoirs to supply energy demands, these impurities are considered objectionable and are removed by various processing schemes. The hydrocarbon gases normally found in natural gas are methane, ethane, propane, butanes, pentanes, and small amounts of hexanes, heptanes, octanes, and heavier gases. Usually the ethane and heavier fractions are removed for additional processing because of their high market value as gasoline blending stock and chemical plant raw feedstock. What usually reaches the transmission line for sale as natural gas is a mixture of methane and ethane with some small percentage of propane. Methane is the largest component (usually 95% to 98%).

Composition of natural gas

Natural gas has been defined as a mixture of hydrocarbon gases and impurities. No one mixture can be referred to as natural gas; each gas stream produced has its own composition. Even two gas wells from the same reservoir may have different compositions. Examples of some typical natural gas streams are provided in table 1.1 to show the range of composition that is naturally produced.

Well stream no. 1 is typical of an associated gas; that is, gas produced with crude oil. Well streams no. 2 and no. 3 are typical low-pressure and high-pressure gases of the nonassociated type. Not only is there a wide variety of natural gas compositions, but also each gas stream produced from a natural gas reservoir can change composition as the reservoir is depleted. Well analyses should be checked periodically, since it may be necessary to change the equipment used for production in order to satisfy the new composition of the gas.

TABLE 1.1
Typical Natural Gas Analyses

Component	Well No. 1 Mol Percent	Well No. 2 Mol Percent	Well No. 3 Mol Percent
Methane	27.52	71.01	91.25
Ethane	16.34	13.09	3.61
Propane	29.18	7.91	1.37
i-Butane	5.37	1.68	0.31
n-Butane	17.18	2.09	0.44
i-Pentane	2.18	1.17	0.16
n-Pentane	1.72	1.22	0.17
Hexane	0.47	1.02	0.27
Heptanes and Heavier	0.04	0.81	2.42
Carbon Dioxide	0.00	0.00	0.00
Hydrogen Sulfide	0.00	0.00	0.00
Nitrogen	0.00	0.00	0.00
Total	100.00	100.00	100.00

Note: Production from many wells will contain small quantities of carbon dioxide, hydrogen sulfide, and nitrogen.

Natural gas is normally thought of as being a mixture of straight-chain, or paraffin, hydrocarbon gases. However, cyclic and aromatic hydrocarbon gases (cyclic compounds) are occasionally found in the mixture. The terms *straight-chain* and *cyclic* refer to the molecular structure. Methane, ethane, propane, butanes, pentanes, and so forth are straight-chain, or paraffinic, compounds. The molecular structure of these and the cyclic compounds are shown for comparison in figure 1.1.

Impurities found in natural gas must be removed because they cause difficulties in

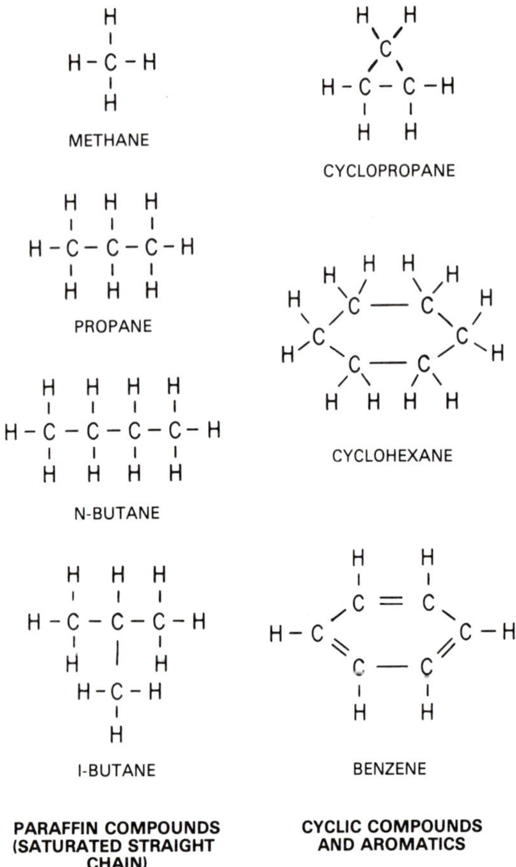

Figure 1.1. Hydrocarbon molecular structures

handling and processing. Components such as hydrogen sulfide, carbon dioxide, mercaptans, water vapor, noncombustible gases (such as nitrogen and helium), pentanes, and the heavier hydrocarbons may cause extremely unreliable and hazardous combustion conditions for the consumer. Removal of hydrogen sulfide, carbon dioxide, and water vapor eliminates the problems of toxicity, corrosion, and hydrate formation in transmission and distribution systems. Market value of some of these impurities may make it beneficial for the producer to remove and market them separately, thereby making the processing of the natural gas more economical. When the pentanes and heavier hydrocarbons are recovered, they provide a good source of blending stocks for the production of gasoline. With the growth of the worldwide market for liquefied petroleum gas (LPG), additional processes for higher recovery of the ethane, propane, and butane fractions, which make up LPG, have been developed. These fractions are valuable for petrochemical industry raw stocks.

Physical properties of natural gas

Since natural gas is a mixture of hydrocarbon compounds of varied types and amounts, the overall or combined physical properties vary. Physical properties of a natural gas are indicators of the behavior of the gas under various processing conditions, and it is therefore important to establish what these properties are. To do this, an analysis of the composition must first be made. Once the composition is known, the various physical properties can be determined by using those of each pure component in the mixture. Physical properties that are most useful to know for natural gas processing are molecular weight, freezing point, boiling point, density, critical temperature, critical pressure, heat of vaporization, and specific heat. Table 1.2 is a tabulation of physical constants of paraffin hydrocarbons and other components of natural gas from the GPSA *Engineering Data Book.*

Terminology

Gauge pressure

Gauge pressure is that pressure indicated by a pressure gauge and is the pressure above or below that of the atmospheric pressure at the point of measurement. Most pressure-measuring devices make use of a Bourdon tube, which is a curved hollow chamber that tends to straighten out as its internal pressure increases. Atmospheric pressure acts on the outside of the Bourdon tube; so actually the movement of the tube is the result of the differential between the internal pressure and

TABLE 1.2
Physical Constants of Paraffin Hydrocarbons and Other Components of Natural Gas

Component	Notes	Methane	Ethane	Propane	Iso-Butane	*n*-Butane	Iso-Pentane	*n*-Pentane	*n*-Hexane	*n*-Heptane	*n*-Octane	*n*-Nonane	*n*-Decane
Molecular Weight		16.043	30.070	44.097	58.124	58.124	72.151	72.151	86.178	100.205	114.232	128.259	142.286
Boiling Point @ 14.696 psia, °F		−258.74(28)	−127.44	−43.73	10.74	31.12	82.11	96.91	155.73	209.16	258.21	303.48	345.49
Freezing Point @ 14.696 psia, °F		−296.45[d]	−297.04[d]	−305.82[d]	−255.28	−217.05	−255.82	−201.51	−139.58	−131.05	−70.17	−64.28	−21.35
Vapor Pressure @ 100 °F, psia		(5000.)	(800.)	188.0	72.39	51.54	20.444	15.575	4.960	1.6201	0.5370	0.1796	0.0609
Density of Liquid @ 60 °F & 14.696 psia													
Relative density @ 60 °F/60 °F	a,b	(0.3)[i]	0.3563[h]	0.5075[h]	0.5630[h]	0.5842[h]	0.6244	0.6311	0.6640	0.6883	0.7070	0.7219	0.7342
API	a,b	(340.)[i]	265.6[h]	147.3[h]	119.8[h]	110.7[h]	95.1	92.7	81.60	74.08	68.64	64.51	61.23
Absolute density, lbm/gal (in vacuum)		(2.5)[i]	2.970[h,x]	4.231[h,x]	4.694[h]	4.870[h]	5.206	5.261	5.536	5.738	5.894	6.019	6.121
Apparent density, lbm/gal (in air)	c	(2.5)[i]	2.960[h]	4.221[h]	4.684[h]	4.861[h]	5.196	5.252	5.527	5.729	5.885	6.010	6.112
Density of Gas @ 60 °F & 14.696 psia													
Relative density, air = 1.00, ideal gas		0.5539	1.0382	1.5225	2.0068	2.0068	2.4911	2.4911	2.9753	3.4596	3.9439	4.4282	4.9125
Lb/M ft^3, ideal gas		42.28	79.24	116.20	153.16	153.16	190.11	190.11	227.07	264.06	301.02	337.95	374.95
Volume @ 60 °F & 14.696 psia													
Liquid, gal/lb-mole		(6.4)[i]	10.12[h]	10.42[h]	12.38[h]	11.93[h]	13.86	13.71	15.59	17.46	19.38	21.31	23.26
ft^3 gas/gal liquid, ideal gas		(59.1)[i]	37.48[h]	36.41[h]	30.65[h]	31.80[h]	27.38	27.67	24.38	21.73	19.58	17.81	16.32
Ratio, gas/liquid in vacuum		(442.)[i]	280.4[h]	272.3[h]	229.3[h]	237.9[h]	204.8	207.0	182.4	162.6	146.5	133.2	122.1
Critical Conditions													
Temperature, °F		−116.68	90.10	206.01	274.96	305.62	369.03	385.6	453.6	512.7	564.10	610.54	651.6
Pressure, psia		667.8	707.8	616.3	529.1	550.7	490.4	488.6	436.9	396.8	360.6	331.8	304.4
Gross Calorific Value, Combustion @ 60 °F													
Btu/lb, liquid	*	——	22,178.[h]	21,499.[h]	21,084.[h]	21,133.[h]	20,884.	20,922.	20,783.	20,680.	20,601.	20,542.	20,494.
Btu/lb, gas	*	23,881.	22,322.	21,662.	21,237.	21,298.	21,037.	21,084.	20,944.	20,840.	20,759.	20,700.	20,651.
Btu/ft^3, ideal gas	*,p,t	1009.7	1768.7	2517.2	3252.6	3262.0	3999.7	4008.7	4756.1	5502.8	6248.9	6996.3	7743.1
Btu/gal liquid	*,t	——	65,889.[h]	90,962.[h]	98,968.[h]	102,918.[h]	108,722.	110,071.	115,055.	118,662.	121,422.	123,642.	125,444.
Volume air to burn one volume, ideal gas		9.54	16.70	23.86	31.02	31.02	38.18	38.18	45.34	52.50	59.65	66.81	73.97
Flammability Limits @ 100 °F & 14.696 psia													
Lower, vol % in air		5.0	2.9	2.1	1.8	1.8	1.4	1.4	1.2	1.0	0.96	0.87[s]	0.78[s]
Upper, vol % in air		15.0	13.0	9.5	8.4	8.4	(8.3)	8.3	7.7	7.0	——	2.9	2.6
Heat of Vaporization @ 14.696 psia													
Btu/lb @ boiling point		219.20	210.39	183.03	157.52	165.63	147.12	153.58	143.94	136.00	129.52	124.17	118.68
Specific Heat @ 60 °F & 14.696 psia													
C_p gas, Btu/(lb · °F), ideal gas		0.5266	0.4080	0.3887	0.3867	0.3951	0.3829	0.3880	0.3857	0.3842	0.3831	0.3822	0.3816
C_v gas, Btu/(lb · °F), ideal gas		0.4027	0.3419	0.3436	0.3526	0.3609	0.3554	0.3605	0.3626	0.3644	0.3657	0.3667	0.3676
$\gamma = C_p/C_v$, ideal gas		1.308	1.193	1.131	1.097	1.095	1.077	1.076	1.064	1.054	1.048	1.042	1.038
C_p liquid, Btu/(lb · °F)		——	0.9256	0.5920	0.566(41)	0.566(41)	0.5353	0.548(41)	0.5332	0.5280	0.5238	0.5220	0.5207
Octane Number													
Motor clear		——	+0.05[f]	97.1	97.6	89.6[j]	90.3	62.6[j]	26.0	0.0	——	——	——
Research clear		——	+1.6[f,j]	+1.8[f,j]	+0.10[f,j]	93.8[j]	92.3	61.7[j]	24.8	0.0	——	——	——
Refractive Index, n, @ 68 °F		——	1.19949[h]	1.28624[h]	——	1.32943[h]	1.35373	1.35748	1.37486	1.38764	1.39743	1.40542	1.41189

SOURCE: GPA Publication 2145[27] — Approval Pending; Abridged

the external. The average atmospheric pressure at sea level is 14.7 psi or 29.92 inches of mercury. Gauge pressure is reported as pounds per square inch gauge (psig).

Absolute pressure

The absolute pressure must be used in all calculations involving volume and pressure relationships. To obtain absolute pressure, the pressure of the atmosphere must be added to the gauge pressure. To be entirely accurate, the atmospheric pressure on the gauge at the time the gauge pressure is read should be determined, but for most calculations the average sea level pressure of 14.7 psi can be added. In higher altitudes, a lower atmospheric pressure should be added. Absolute pressure is reported as pounds per square inch absolute (psia).

Vapor pressure

When a liquid evaporates into a space of limited dimensions, the space becomes filled with the vapor that is formed. As vaporization proceeds, the number of molecules in the vapor state increases and causes an increase in the pressure exerted by the vapor. The pressure exerted by a gas, or vapor, is due to the impacts of its component molecules against the confining surfaces. Since the original liquid surface forms one of the walls confining the vapor, it receives a continual series of impacts by the molecules in the vapor state. The pressure exerted by the vapor is dependent upon the number of such impacts. However, when one of the gaseous molecules strikes the liquid surface, it comes under the influence of the attractive forces of the densely aggregated liquid molecules, is held there, and forms a part of the liquid once more. This phenomenon, the reverse of vaporization, is known as *condensation*. The rate of condensation is determined by the number of molecules striking the liquid surface per unit of time, which is in turn determined by the pressure of the vapor.

When a liquid evaporates into a limited space, two opposing processes are in operation. The process of vaporization tends to change the liquid into the gaseous state; the process of condensation tends to change the gas formed by vaporization back into the liquid state. The rate of condensation is increased as vaporization proceeds, thus increasing the pressure of the vapor. Ultimately, if sufficient liquid is present, the processes will continue until the rate of condensation equals the rate of vaporization. When this condition is reached, a dynamic equilibrium is established. Since the formation of new vapor is compensated by condensation, the pressure of the vapor will remain unchanged. If the pressure of the vapor is changed in either direction from its equilibrium value, it will adjust itself and return to the equilibrium condition. The pressure exerted by the vapor at such equilibrium conditions is termed the *vapor pressure* of the liquid.

All materials exhibit a definite vapor pressure of greater or lesser degree at any temperature above absolute zero. In general, as molecular weight increases, the vapor pressure decreases. The magnitude of the vapor pressure is not dependent upon the amount of liquid present, as long as some liquid is present. Neither is the magnitude dependent upon the amount of liquid surface area present. Vapor pressure is dependent entirely on the magnitudes of the maximum potential energies of attraction, which must be overcome in vaporization.

Partial pressure and pure component volume

In a mixture of different gases, the molecules of each component gas are distributed throughout the entire volume of the containing vessel, and the molecules of each component gas contribute by their impacts to the total pressure exerted by the entire mixture. The total pressure is equal to the sum of the pressures exerted by the molecules of each component gas. These statements apply to all gases, whether or not their behavior is ideal. In a mixture of ideal gases, the molecules of each component gas behave independently, as though they alone were present in the container. Before considering the actual behavior of the gaseous mixtures, it is necessary to define two terms commonly used, namely, *partial pressure* and *pure component volume.* The *partial pressure* of a component gas in a mixture of gases is the pressure that would be exerted by that gas if the same number of molecules of that gas were present alone in the same volume and at the same temperature as the mixture. The *pure component volume* of that gas is the volume that it would occupy if it existed by itself in the same container at the same conditions of temperature and pressure.

The kinetic theory of the constitution of gases is based on the phenomenon that many properties of gaseous mixtures are additive. The additive nature of partial pressures is expressed by Dalton's law, which states that the total pressure exerted by a gaseous mixture is equal to the sum of the partial pressures.

$$P = p_a + p_b + p_c + p_d + \dots.$$

where P is the total pressure and p_a, p_b, p_c, and so forth are the partial pressures of the component gases.

Similarly, the additive nature of pure component volumes is given by the law of Amagat and Leduc, which states that the total volume occupied by a gaseous mixture is equal to the sum of the pure component volumes.

$$V = v_a + v_b + v_c + \dots.$$

where V is the total volume of the mixture and v_a, v_b, v_c, and so forth are the pure component volumes of the component gases.

Standard conditions

An arbitrarily specified standard pressure is established by various states and countries as a *pressure base.* This furnishes a convenient standard that can be used to compare different quantities of gas when they are expressed in terms of volume. Natural gas is produced and sold by volume in the United States and several other countries; therefore, before the volumes of various gases can be compared, the pressure base of each volume must be known. All volume measurements are mathematically converted to the established pressure base before comparison. Some states in the United States have adopted 14.4 psi as their pressure base. Others have adopted pressure bases of 14.7 psi and 15.2 psi. Coupled with the pressure base should be a *standard temperature,* since volume is affected by temperature as well as pressure. All measurement in the United States is referred to a standard temperature of 60°F. Consequently, specification of the pressure base is important in making volume measurement comparisons.

Gas density and specific gravity

The density of a gas is usually expressed as the weight in pounds per cubic foot. Unless specified otherwise, the volumes are at standard conditions of 60°F and 14.7 psia. On this basis, air has a normal density of 0.0763 lb/cu ft. Specific gravity is the ratio of the density of a gas to the density of air at the same standard conditions of temperature and pressure. Specific gravity is an important factor in gas measurement.

Liquid density and specific gravity

The density of a liquid is usually expressed as the weight in pounds per cubic foot, similar to the density of gas. However, the specific gravity of liquids is different. In the case of liquids (including liquefied natural gas, liquid propane, etc.), specific gravity is the ratio of a liquid's density to that of water at 60°F.

Another gravity term used with hydrocarbon liquids is API gravity. A special gravity scale for expressing gravity of petroleum products has been adopted by the American

Petroleum Institute. The scale is defined as follows:

$$\text{Degrees API} = \frac{141.5}{G} - 131.5$$

G is the liquid's specific gravity at 60°F referenced to that of water at 60°F. Thus a liquid that has the same density as water at 60°F, that is, a specific gravity of 1.0, will have an API gravity of 10°API. The gravity of a liquid in degrees API is determined by its density at 60°F. Readings of API-graduated hydrometers at temperatures other than 60°F must be corrected for temperature to give the value at 60°F.

Temperatures

The Fahrenheit temperature scale is the one used almost universally in the gas industry. However, in most other industries throughout the world, the Celsius scale is used. This scale is a part of the SI system of units (Système International, or international system of units). It has its zero established at the freezing point of water and its 100 mark at the atmospheric (sea level) boiling point of water. These are the easiest reference points for scientists and were used as calibration points for early thermometers. The Fahrenheit scale was devised by using the same reference points, but the freezing point of water was designated as 32 degrees and the boiling point of water as 212 degrees.

Absolute temperature came into use when scientists began expanding their investigations of matter and found that correlations could be established using temperature as a variable, provided that the temperature was expressed in absolute terms—that is, in reference to absolute zero. By definition, *absolute zero* is the temperature at which there is very little or no movement taking place by the atoms within matter. In the Celsius scale, this is 273.1 degrees below 0°C. It is 0 kelvin on the kelvin scale. In other words, 0°C corresponds to 273.1 k. (Note that *kelvin* is not capitalized and that the units are not expressed in degrees.) For the Fahrenheit scale, absolute zero is 460 degrees below 0°F. This absolute scale is referred to as the Rankine scale. Thus,

$$XC = (X + 273.1)\text{k}$$
$$XF = (X + 460)\text{R}$$

Figure 1.2 shows a comparison of the absolute, Celsius, and Fahrenheit scales. In working with gas laws, care must be taken to use the absolute scale for temperature and to use consistent units for the expressions of both the variable and the constant terms.

Early temperature-measuring instruments were limited in their ranges by the materials available. Today a wide choice is available, ranging from liquid-filled thermometers to thermocouples to optical pyrometers.

Mass and weight

All matter has mass and weight even though they may be small. The gases that make up natural gas vary in their mass and weight. Since the word *weight* has become virtually synonymous with *mass* in engineering literature, it has become common practice to use the weight of a material rather than its mass as the measure of quantity. Mass and weight are equal only at sea level, but the variation of weight at different points on the earth's surface is negligible in ordinary engineering work.

Molecular weight

Natural gas is a mixture of several different gases that have different characteristics, one of which is structure (fig. 1.1). The structure of a gas and its molecular weight are related. The structure, expressed as a formula, indicates the relative numbers and kinds of atoms that unite to form the gas molecule. For example, the formula for methane, CH_4, indicates that carbon and hydrogen are present in the compound in a 1 to 4 ratio. By

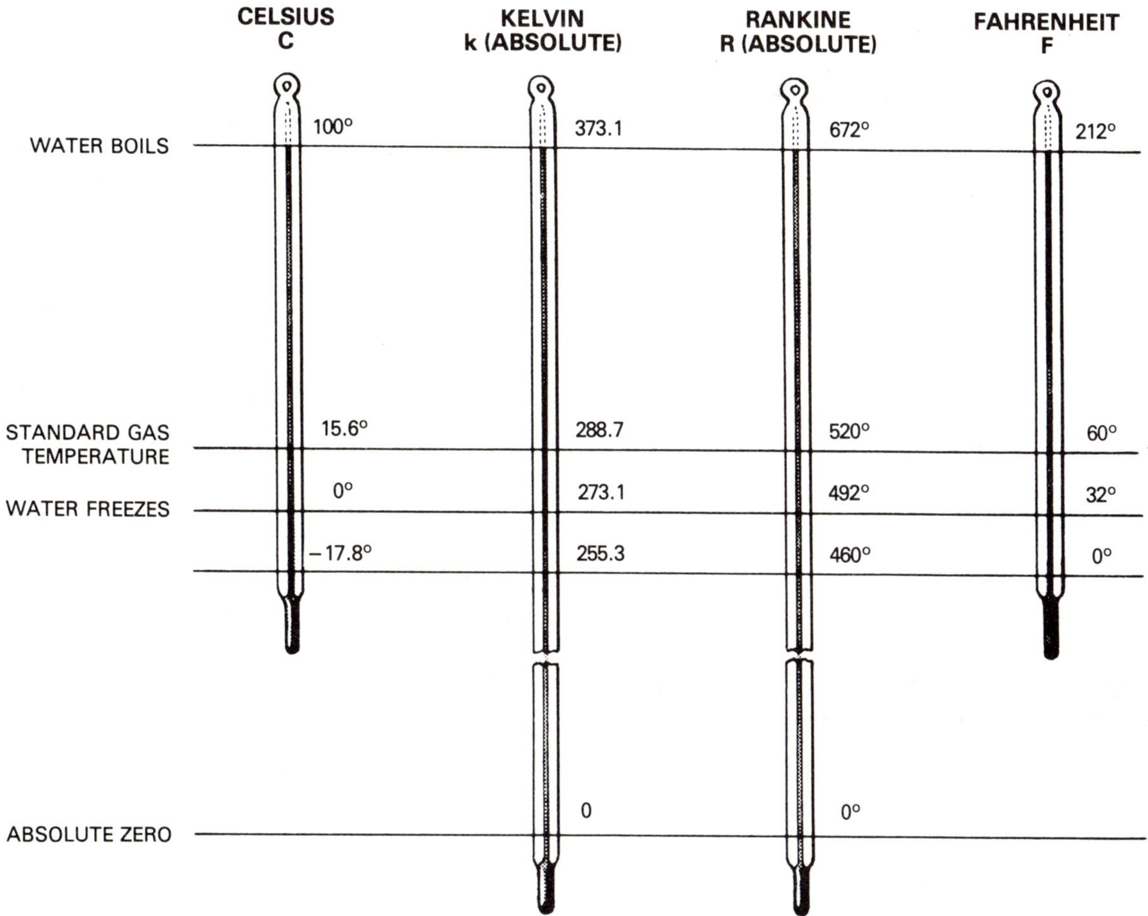

Figure 1.2. Comparison of absolute, Celsius, and Fahrenheit temperature scales

taking the atomic weight of carbon and adding to it four times the weight of hydrogen, the molecular weight of methane can be determined—16.043, as shown in table 1.2.

Pound-atom and pound-mol

The mass of a given element in pounds, equal numerically to its atomic weight, is termed a *pound-atom.* Similarly, the mass of a molecule expressed in pounds numerically equal to its molecular weight is termed a *pound-mol.* A pound-mol of methane (CH_4) weighs 16.043 pounds and contains a specific number of methane molecules. That same number of molecules of ethane would make a pound-mol of ethane (C_2H_6) but would weigh 30.070 pounds.

Mol fraction

Another term that is quite useful in natural gas operations is *mol fraction.* In a mixture of two hydrocarbon gases such as methane and ethane, there are two kinds of molecules. The number of methane molecules divided by the sum of methane and ethane molecules would represent the mol fraction of methane in the gaseous mixture. The mol fraction times 100 is the *mol percent.* For all practical purposes, the mol percent and the volume percent are the same. This relationship holds only for gases and does not apply to liquid systems.

Critical properties

Critical temperature is the temperature above which a pure hydrocarbon gas cannot be liquefied no matter how much pressure is applied. For example, the critical temperature for methane is −116.6°F (table 1.2). If the temperature of methane gas is above this, it cannot be liquefied regardless of how much pressure is applied to it.

Table 1.2 shows another critical property of gases, critical pressure. The *critical pressure* is the pressure required to liquefy a gas at its critical temperature. In the previous example, methane's critical temperature was given as −116.6°F. At this temperature, a pressure of 667.8 psia would be needed in order to liquefy the gas.

Ideal gas laws

The ideal gas law has been empirically developed from extensive experimental investigations. In fact, the definition of the absolute scale of temperature is based on this relationship.

$$Pv = RT \tag{1}$$

or

$$PV = nRT \tag{2}$$

where

R = a proportionality factor
T = absolute temperature
v = volume of 1 mol of gas
n = number of mols of gas
V = volume of n mols of gas
P = absolute pressure.

When equation (1) is rearranged,

$$R = \frac{Pv}{\mathrm{T}} \tag{3}$$

Assuming from Avogadro's hypothesis that equimolar quantities of all gases occupy the same volume at the same conditions of temperature and pressure, it follows from equation (3) that the gas law factor R is a universal constant. The Avogadro hypothesis and the ideal gas law have been experimentally shown to approach perfect validity for all gases under conditions of extreme rarefaction—that is, where the number of molecules per unit volume is very small. The constant R may be evaluated from a single measurement of the volume occupied by a known molar quantity of any gas at a known temperature and at a known *reduced* pressure.

In using the gas law equations, great care must be taken to employ consistent units throughout any computation. Pressure and temperature must be in absolute units such as psia and degrees Rankine. The numerical value of the gas constant R has been carefully determined and may be expressed in any desired energy units (table 1.3).

When substances exist in the gaseous state, two general types of problems arise in determining the relationships among weight, pressure, temperature, and volume. The first type involves only the last three variables—pressure, temperature, and volume. For example, given a specified volume at a

TABLE 1.3
Values of *R* Corresponding to Various Systems of Units

Units of Pressure	Units of Volume	*R*
Per gram-mol (temperature, kelvin)		
Atmospheres	Cubic centimetres	82.060
Per pound-mol (temperature, degrees Rankine)		
Pounds per square inch	Cubic inches	18,510.000
Pounds per square inch	Cubic feet	10.710
Atmospheres	Cubic feet	0.729

specified temperature and pressure, the volume at any other specified pressure and temperature combination can be calculated. For such calculations, the weight of gas is not required. The second, more general type of problem involves the weight of gas. For example, a specified weight of a substance exists in the gaseous state under certain conditions, two of which are specified and the third of which is to be calculated. Or conversely, the weight of a given quantity of gas existing at specified conditions of temperature, pressure, and volume can be calculated.

Problems of the first type may be solved by means of the proportionality indicated by the gas law. Equation (2) may be applied to n mols of gas at conditions P_1, V_1, T_1, and at conditions P_2, V_2, and T_2.

$$P_1 V_1 = nRT_1$$

$$P_2 V_2 = nRT_2$$

Combining the two equations,

$$\frac{P_1 V_1}{P_2 V_2} = \frac{T_1}{T_2} \tag{4}$$

This equation may be applied directly to any quantity of gas. If the three conditions of state 1 are known, any of those of state 2 may be calculated to correspond to specified values of the other two. Any units of absolute pressure, volume, and absolute temperature may be used, the only requirement being that the units in both initial and final states be the same.

Problems of the second type, in which both weights and volumes of gases are involved, can be solved directly by use of the equation $PV = nRT$. With weights expressed in molal units, the equation may be solved for any one of the four variables if the other three are known. However, the value of the constant R must be expressed in units to correspond to those used in expressing the four variable quantities.

The ideal gas law permits satisfactory calculations of pressure-volume-temperature relationships at low pressures when the volumes per mol are relatively large, the distances between molecules are great, and the temperatures are relatively high. However, under conditions of small molar volumes and high pressures, the errors in assuming ideal gas behavior may be as great as 500 percent. Several hundred equations have been proposed by various scientists to express the correct PVT relationship of gases, but none has been found to be universally satisfactory. These equations are referred to as equations of state. All are cumbersome to use, so engineers have developed a substitute equation of state using a compressibility factor that compensates for the deviations from the ideal gas law.

This equation of state may be written as follows:

$$PV = ZnRT$$

Z in this case is the compressibility factor; it is a function of temperature, pressure, and nature of the gas. Compressibility factors for a large number of gases have been determined experimentally. They have also been tabulated in different ways over a wide range of temperatures and pressures. Usually, the compressibility factor is plotted or tabulated as a function of pressure and temperature for the specific gravity of a given gas. The most useful relationship for compressibility is a plot of the compressibility factor as a function of reduced pressure and reduced temperature.

In order to use a chart such as that in figure 1.3, it is necessary to know the critical temperature and the critical pressure of the gas in question. The reduced properties of a gas are calculated as follows:

$$T_r = \frac{T}{T_c} \qquad P_r = \frac{P}{P_c}$$

where the subscript r indicates the reduced, and c the critical, properties. Both temperature and pressure must be expressed in absolute units.

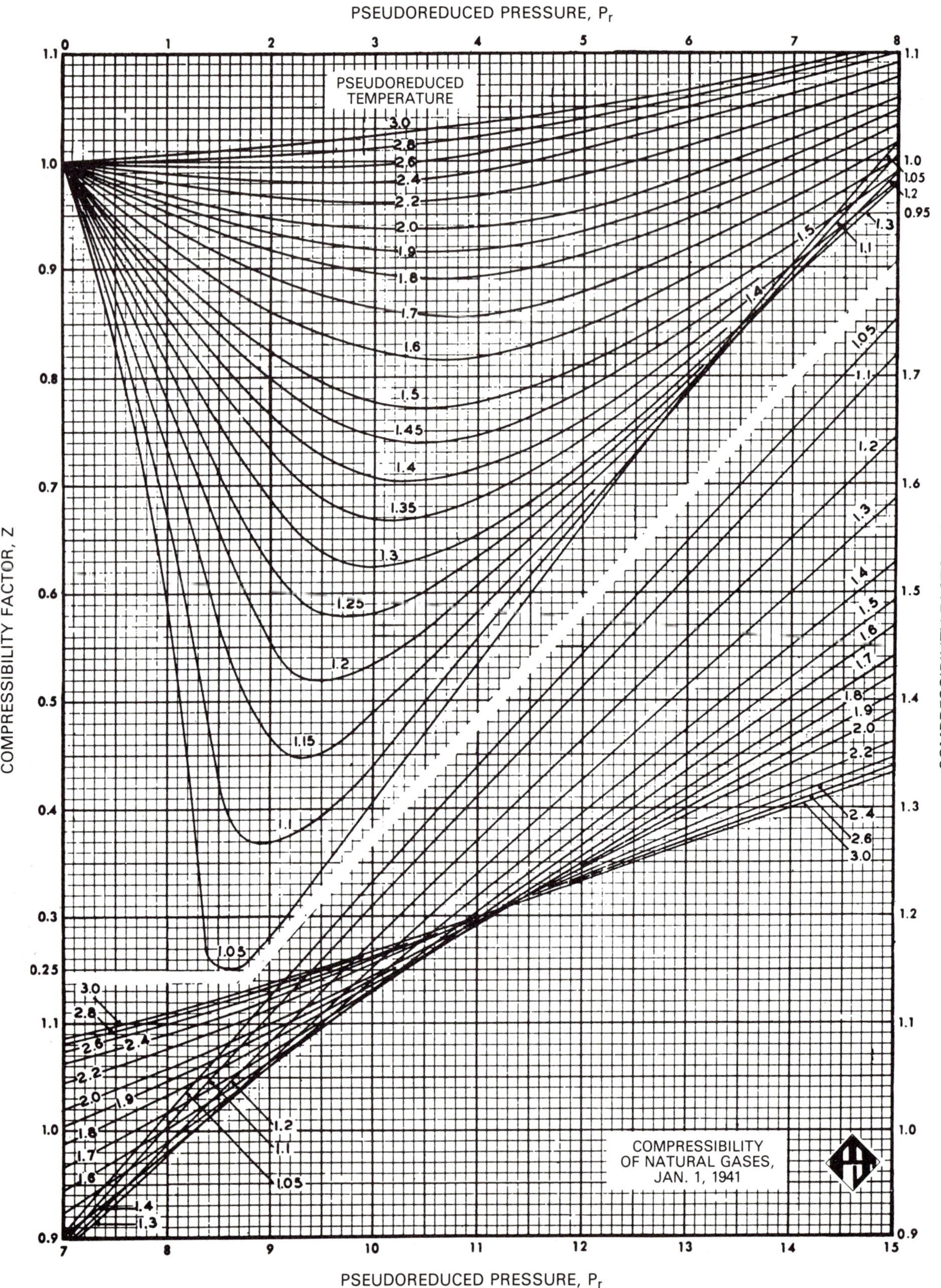

Figure 1.3. Compressibility factors for natural gas

For gas mixtures such as those in natural gas, critical properties are not easily found in books and tables; so general use is made of what are termed pseudocritical properties. Pseudocritical properties are calculated by using the mol fraction and the critical properties of each pure component (table 1.4). In this table, the gas used is a mixture of methane, ethane, and propane, with the composition in mol fractions as given.

Using the pseudocritical properties calculated in the table and figure 1.3, the compressibility factor for this gas at 100°F and 300 psig can be determined as follows:

$$T = 100°\text{F} + 460° = 560°\text{R}$$

$$T_r = \frac{T}{T_c'} = \frac{560°}{647.2°} = 0.868$$

$$P = 300 \text{ psig} + 14.7 = 314.7 \text{ psia}$$

$$P_r = \frac{P'}{P_c'} = \frac{314.7}{627.1} = 0.502$$

Using $T_r = 0.868$ and $P_r = 0.502$, $Z = 0.64$.

Sometimes another term is used by engineers to compensate for the deviation from the ideal gas law. This term is referred to as the *supercompressibility factor* (F_{pv}) and is used primarily for high-pressure calculations. There is a mathematical relationship between supercompressibility and compressibility, as follows:

$$Z = \frac{1}{(F_{pv})^2}$$

Substituting the supercompressibility factor for the compressibility factor, the ideal gas law equation can now be written—

$$PV = \frac{nRT}{(F_{pv})^2}$$

Tables of supercompressibilities for natural gas are available from a number of sources and can be readily used with the above equation to predict PVT relationships.

Equilibrium concepts

In handling natural gas mixtures, the engineer must be able to predict the exact relationship among all the components at any given temperature and pressure. Not only does he have to know whether he has a vapor, liquid, or combination of both, but he must also be able to predict the distribution of each component in each phase. For this reason, engineers have devised the *equilibrium constant,* or K value. The equilibrium constant is known as the *vapor-liquid equilibrium ratio* and is designated as Y_i/X_i, where Y_i is the mol fraction of component i in the vapor phase and X_i is the mol fraction of component i in the liquid phase. The K factor is a function of temperature, pressure, and the composition of a particular system. K values most widely used are those which have been developed by GPA. In order to use the GPA equilibrium constants properly, the engineer must be able to predict the convergence pressure of the natural gas stream being investigated.

Convergence pressure (P_k) is the pressure at a given temperature when the vapor-liquid

TABLE 1.4
Calculation of Pseudocritical Properties

Component	Mol Fraction *(N)*	T_c *(R)*	NT_c	P_c *(psia)*	NP_c
Methane (CH_4)	0.015	343	5.1	668	10.1
Ethane (C_2H_6)	0.120	550	66.0	708	85.0
Propane (C_3H_8)	0.865	666	576.1	616	532.0
			$T_c' = 647.2$		$P_c' = 627.1$

equilibrium ratios for the various components in a system become or tend to become equal to unity (1.0). Figures 1.4, 1.5, and 1.6 are K charts for a convergence pressure of 2,000 psia for methane, ethane, and isobutane. Note that all the curves for various temperatures converge at a K value of 1.0 and a pressure of 2,000 psia.

Convergence pressure can be determined by several methods, but the most prevalent one is that based on the ideas of Hadden and Winn, which is discussed in the Gas Processors Suppliers Association *Engineering Data Book.* It involves the selection of a pseudobinary system consisting of (1) the lightest component present in an amount of at least 0.1 mol fraction in the liquid phase and (2) the remaining components of the liquid that are heavier.

A problem lies with the heavier components, because selection of one component that is representative of the group must be made. A suitable solution is to use the weighted average critical pressure and temperature of the combined heavier components. The usual shortcut is to calculate the pseudocritical pressure of the heavier component portion and then pick the single component, such as butane, pentane, and so forth, that has a critical temperature closest to the pseudocritical temperature of the heavier components. Using the lightest component and the component selected above as representative of the heavier components, a binary system is chosen. Figure 1.7 is used to determine the approximate convergence pressure versus the system temperature. With this approximate value for the convergence pressure, the selection of the series of K values having a convergence pressure equal to or greater than the calculated convergence pressure is possible.

Figure 1.7 is a plot of the convergence pressure for various binary systems as a function of temperature and is very useful to the engineer in determining the exact vapor-liquid relationship for any given system or any temperature and pressure.

In actual practice, an *assumed* convergence pressure is used to establish *estimated K* values and an equilibrium calculation performed to arrive at an *estimated* amount and composition of the liquid phase at given operating conditions for a given hydrocarbon system. This practice provides the basis for selecting the pseudobinary system referred to above. If the convergence pressure calculated from this binary system approximates the assumed convergence pressure, K values so established may be used with reliance in equilibrium calculations. If the assumed and calculated convergence pressures are far apart, the calculation must be repeated, using an assumed pressure more closely approaching that obtained in the initial calculation.

K values have been established for a number of convergence pressures, specifically 800, 1,000, 1,500, 2,000, 3,000, 5,000, and 10,000 psia. To obtain the K value for a given component, it is necessary only to locate the intersection of the operating pressure and temperature curves on the appropriate convergence pressure chart. For example, see figures 1.4, 1.5, and 1.6, where the K values for methane, ethane, and isobutane have been determined at 3.20, 0.41, and 0.045, respectively, for an operating pressure of 600 psia and temperature of $-20°F$ on a 2,000 psia convergence pressure K chart.

Before proceeding with an illustrative problem of vaporization equilibrium calculations (flash calculations), two additional terms should be defined. These are *bubble point* and *dew point.* Bubble point is the temperature at which the first bubble forms in a liquid at a given pressure. In other words, it is the temperature at which the hydrocarbon mixture begins to vaporize at a given pressure. At the bubble point, the sum of the K_iN_i equals 1.0, where K_i is the equilibrium vaporization constant and N_i is the mol fraction for each component. Dew point is the temperature at which the first drop of liquid forms in the system at a given pressure. It is the

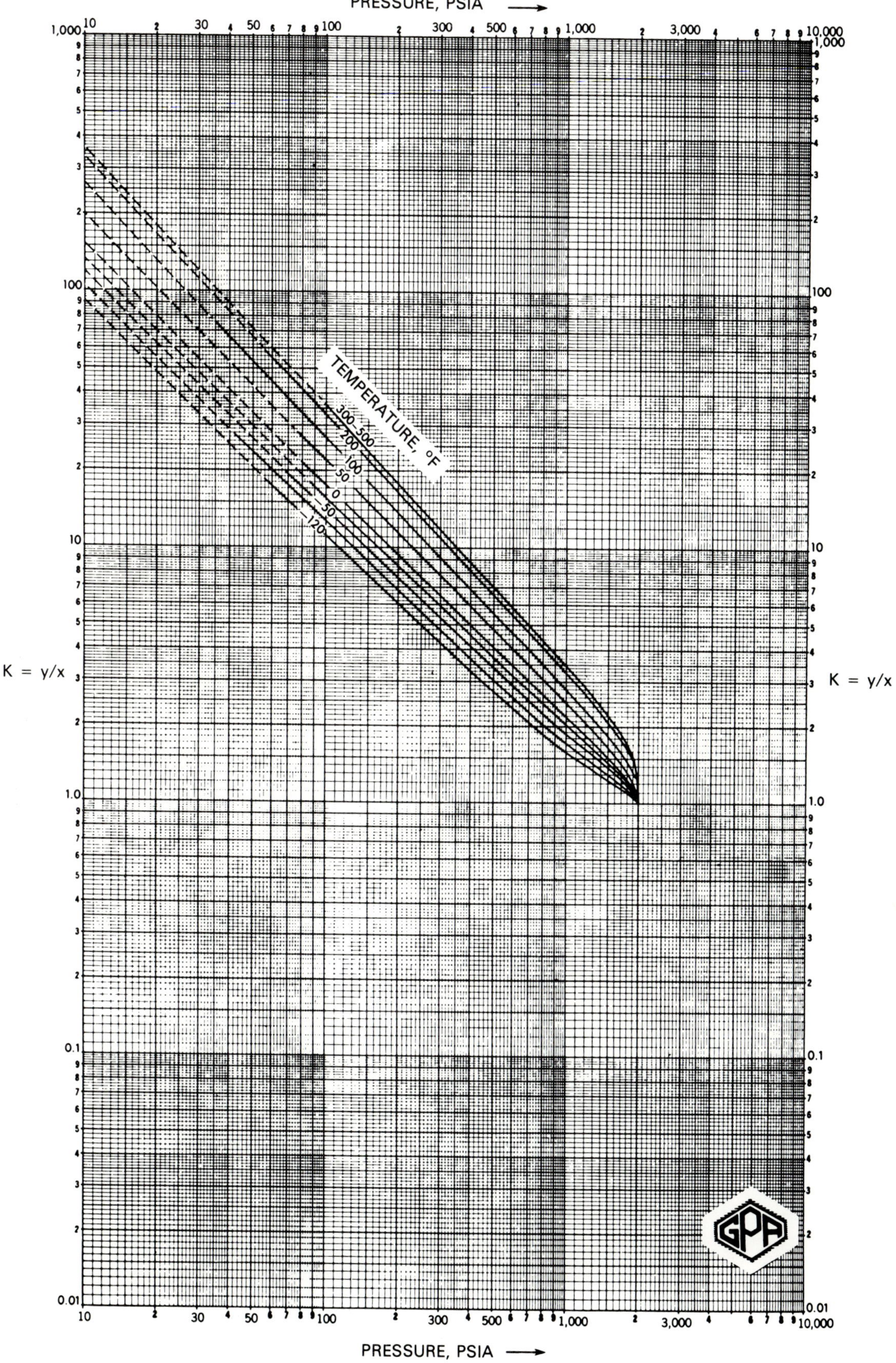

Figure 1.4. Methane convergence pressure at 2,000 psia

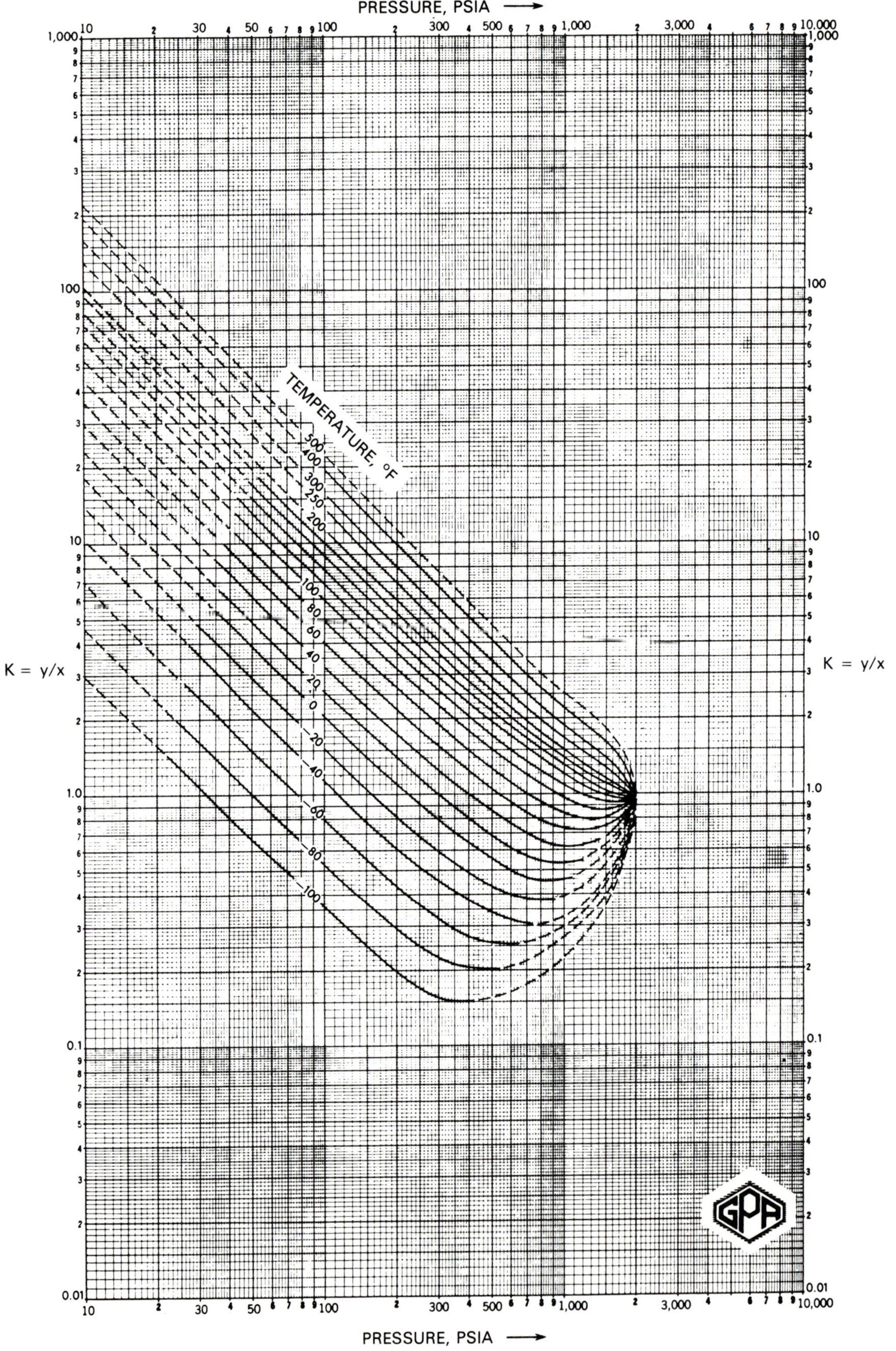

Figure 1.5. Ethane convergence pressure at 2,000 psia

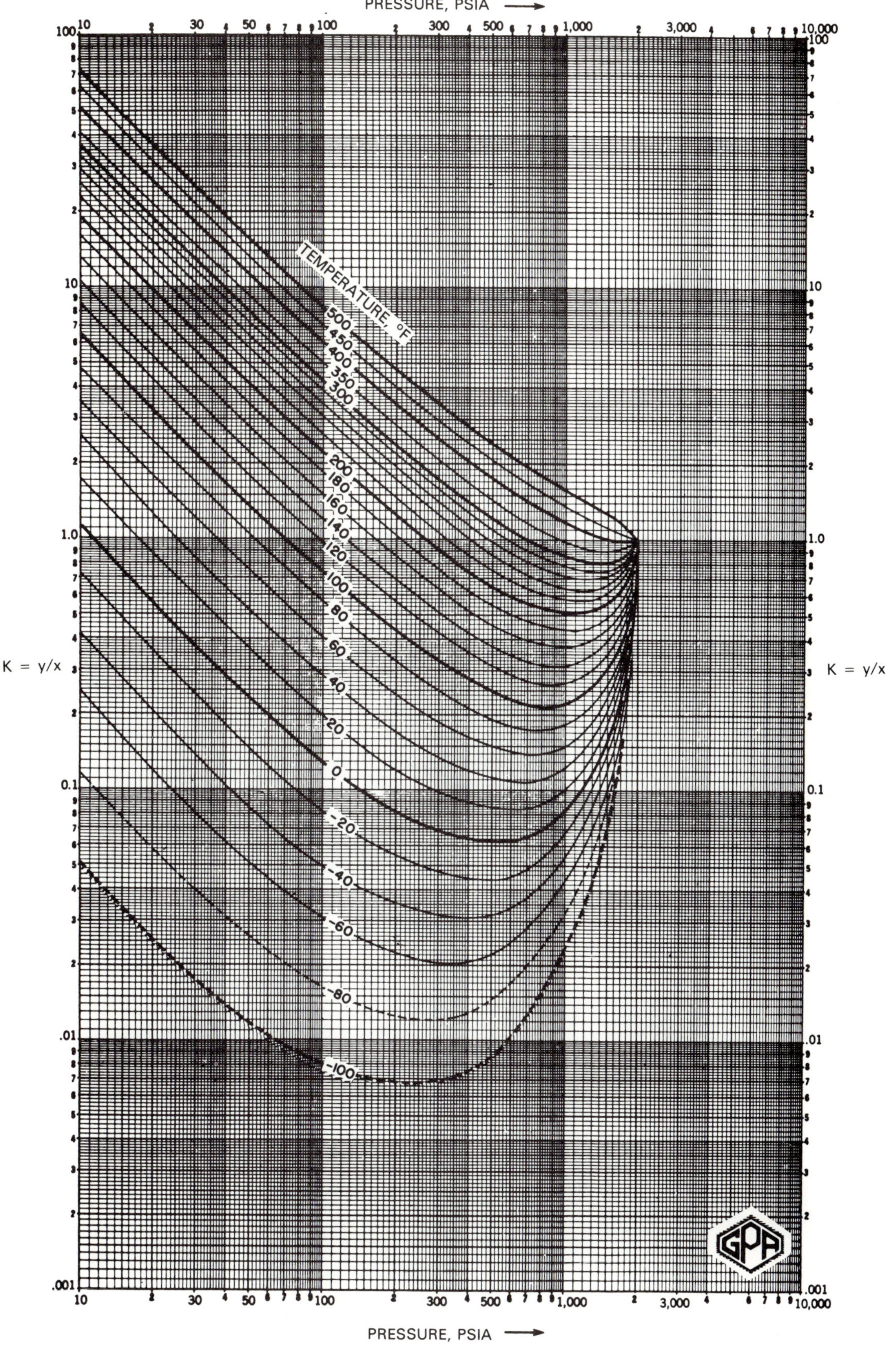

Figure 1.6. Isobutane convergence pressure at 2,000 psia

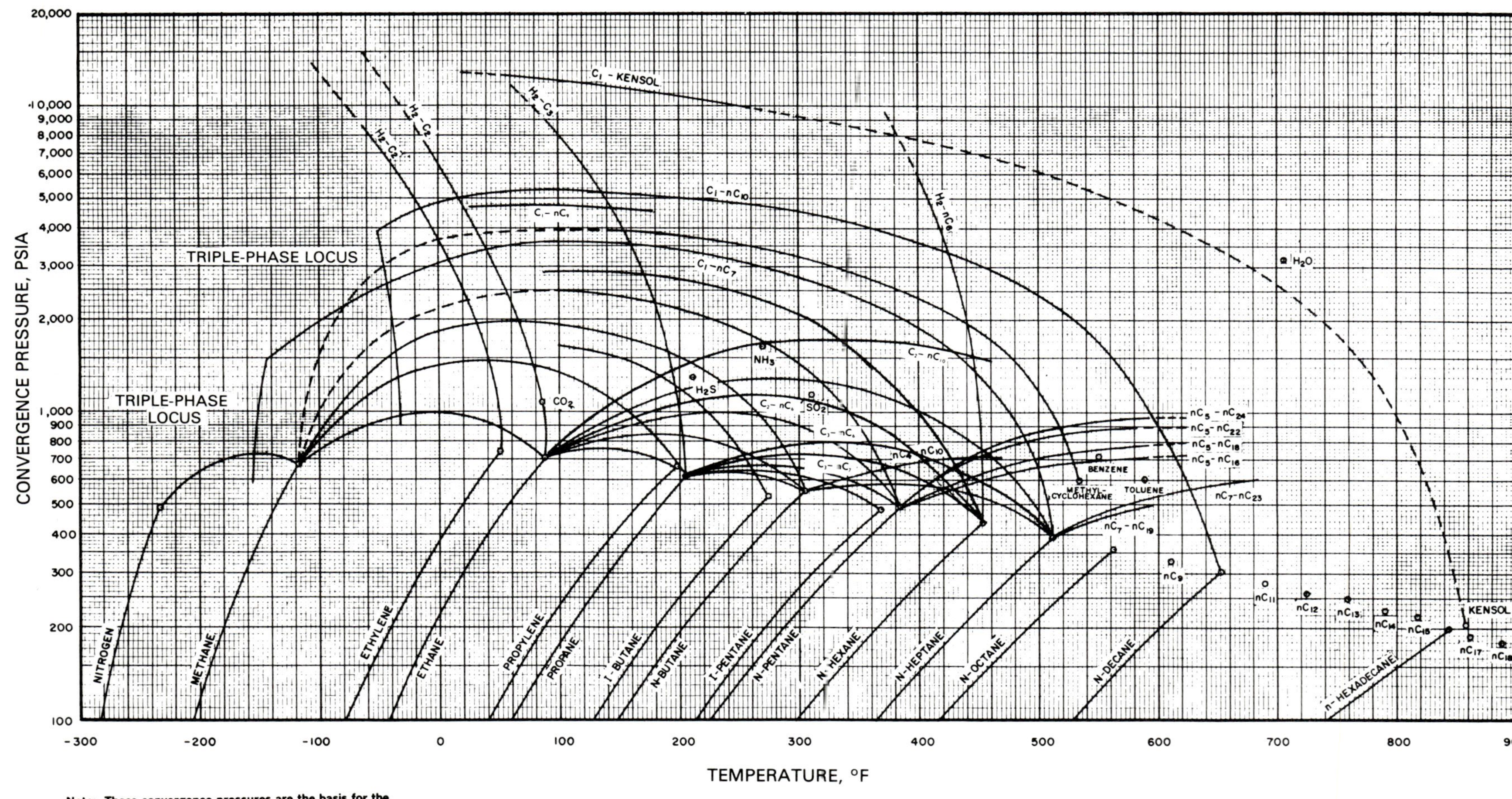

Note: These convergence pressures are the basis for the convergence charts. Later data do not necessarily agree.

Figure 1.7. Convergence pressures for hydrocarbons

temperature at which condensation takes place at a given pressure. At the dew point, the sum of N_i/K_i equals 1.0.

It is wise for the engineer to check a system for its bubble point or dew point before proceeding with the vaporization equilibrium calculation, as this will indicate whether or not there is a single-phase system (all liquid or all vapor) or a two-phase system (vapor and liquid). Flash calculations have no application where a single-phase system exists.

The following problem has been taken from the Gas Processors Suppliers Association *Engineering Data Book.* It is included in order to show the general procedure followed in designing equipment to process natural gas streams and predicting the recovery of hydrocarbon liquids. The equilibrium constants (*Ks*) are those developed by the GPA and published in the above reference. Only a few selected charts, figures 1.4, 1.5, and 1.6, are reproduced to show how the *K* values are determined for the example problem.

For reference, the variables and flash equations are defined as follows:

N_i = mol fraction of component *i* in the total mixture or system

X_i = mol fraction of component *i* in the liquid phase

Y_i = mol fraction of component *i* in the vapor phase

K_i = Y_i/X_i at thermodynamic equilibrium

L = mol fraction of liquid in the total mixture

V = mol fraction of vapor in the total mixture

Σ = summation of calculated terms indicated, that is, ΣX_i and ΣY_i.

A situation of reproducible steady state conditions in a piece of equipment does not necessarily imply that classical thermodynamic equilibrium exists. If the steady composition differs from that for equilibrium, the reason can be the result of time-limited mass transfer and diffusion rates. This warning is made because it is not at all unusual for flow rates through equipment to be so high that equilibrium is not attained or even closely approached. In such cases, equilibrium flash calculations as described here fail to predict conditions in the system accurately, and the *K* values are suspect for this failure—when in fact they are not at fault.

By using the relationships $K = Y/X$ and $L + V = 1.0$ and by writing a material balance for each component in the liquid, vapor, and total mixture, one may derive the flash equation in various forms. A common one is—

$$\Sigma X_i = \Sigma \frac{N_i}{L + VK_i} = 1.0 \qquad (1)$$

Other useful versions may be written as—

$$L = \Sigma \frac{N_i}{1 + (V/L)K_i} \text{ or} \qquad (2)$$

$$\Sigma Y_i = \Sigma \frac{K_i N_i}{L + VK_i}$$

At the phase boundary of the bubble point ($L = 1.0$) and the dew point ($V = 1.0$), these equations reduce to—

$$\Sigma K_i N_i = 1.0 \text{ (bubble point)}$$
$$\Sigma N_i/K_i = 1.0 \text{ (dew point)}$$

These equations are often helpful for preliminary calculations when the phase condition of a system at a given pressure and temperature is in doubt. If the $\Sigma K_i N_i$ is less than 1.0, the system is all liquid and compressed above its bubble point. If $\Sigma N_i/K_i$ is less than 1.0, the sample is all vapor and may be either above its upper dew point or below its lower dew point. The $\Sigma K_i N_i$ and $\Sigma N_i/K_i$ cannot both be less than 1.0 under the same conditions.

Problem

A typical high-pressure separator gas is used for feed to a natural gas liquefaction plant, and a preliminary step in the process involves cooling to −20°F at 600 psia to knock

TABLE 1.5
Flash Calculation at 600 psia and − 20°F

Column	1	2	3	4	5	6	7	8	9	10	11	12	13
			Trial values of L			Final L = .030			Methane free basis				
Component	Feed-gas compo-sition N	P_k = 2000 K	L = .02 $\frac{N_i}{L+VK_i}$	L = .06 $\frac{N_i}{L+VK_i}$	L = .04 $\frac{N_i}{L+VK_i}$	$L+VK_i$	Liquid $X_i = \frac{N_i}{L+VK_i}$	Vapor Y_i	MW	T_c (°R)	P_c (psia)	lb	wt fr
C_1	0.9010	3.2	0.28549	0.29368	0.28952	3.13400	0.28749	0.91997					
CO_2	0.0106	1.14	0.00932	0.00937	0.00934	1.13580	0.00933	0.01064	44.01	547.5	1071.0	0.4106	0.0093
C_2	0.0499	0.41	0.11830	0.11203	0.11508	0.42770	0.11667	0.04783	30.07	549.8	707.8	3.5083	0.0798
C_3	0.0187	0.097	0.16252	0.12369	0.14047	0.12409	0.15070	0.01462	44.10	665.7	616.3	6.6459	0.1511
iC_4	0.0065	0.038	0.11356	0.06791	0.08499	0.06686	0.09722	0.00369	58.12	734.6	529.1	5.6504	0.1284
nC_4	0.0045	0.024	0.10340	0.05451	0.07138	0.05328	0.08446	0.00203	58.12	765.3	550.7	4.9088	0.1116
iC_5	0.0017	0.0088	0.05939	0.02490	0.03509	0.03854	0.04411	0.00039	72.15	828.7	490.4	3.1825	0.0723
nC_5	0.0019	0.0062	0.07286	0.02886	0.04135	0.03601	0.05276	0.00033	72.15	845.3	488.6	3.8066	0.0865
C_6	0.0029	0.0019	0.13265	0.04694	0.06934		0.09107	0.00017	86.18	913.3	436.9	7.8484	0.1784
*C_{7+}	0.0023	0.00066	0.11140	0.03794	0.05661	0.03064	0.07506	0.00005	107.0	998.1	378.7	8.0314	0.1826
		Σ =	1.16889	0.79983	0.91317		1.0089	0.99972		807.7 −459.7 348.0°F	513.9	43.9929	1.0000

*Average nC_7 + nC_8 properties

SOURCE: Gas Processors Suppliers Association, *Engineering Data Book*, 1981.

out heavier hydrocarbons prior to cooling to lower temperatures when these components would freeze out as solids. What is the composition of the feed gas after the preliminary cooling?

Solution

The feed gas composition is shown in table 1.5, column 1. P_k is first estimated to be 2,000 psia, and *Ks* are obtained from 2,000-psia *K* charts and shown in column 2. Then the flash equation (1) is solved for three estimated values of L as shown in columns 3, 4, and 5. By interpolating, the correct value of L for $\Sigma X_i = 1.0$ is determined to be 0.030, as shown in columns 6 and 7. The gas composition is then calculated, using $Y_i = K_i X_i$, as in column 8. This is the composition of the gas leaving the cooler.

To check the convergence pressure assumption, the components heavier than methane in the liquid are converted to a weight-fraction basis as shown in columns 12 and 13, using the molecular weights in column 9. Using this composition (column 13) of the pseudoheavy component, the weighted average critical temperature and pressure is calculated using the T_{ci} and P_{ci} in columns 10 and 11; the results are shown at the bottom of these columns. By referring to table 1.2, the calculated T_c' of 348°F is seen to fall between n-butane and n-pentane, and the calculated P_c' falls in line with the other paraffin hydrocarbons. Referring to figure 1.7, the critical pressure for the pseudobinary, or the convergence pressure, may be calculated at −20°F without actually sketching in a new locus, as follows:

$$P_{kx}' = \frac{(T_{cx}' - T_{c4}')}{(T_{c5}' - T_{c4})} \times (P_{k5} - P_{k4}) + P_{k4}$$

P_{k4} and P_{k5} are read from figure 1.7 at the points where the −20°F line intersects the critical loci curves for methane n-butane (1,700 psia) and methane n-pentane (2,000 psia). T_{cx}' refers to the pseudobinary and comes from column 10, table 1.5 (807.8°R). T_{c4} and T_{c5} are critical temperatures of n-butane and n-pentane respectively, expressed in degrees R and obtained from table 1.2.

$$P_k' = \frac{(807.8 - 765.6)}{(845.5 - 765.6)} \times (2{,}000 - 1{,}700) + 1{,}700$$

$$P_k' = \frac{42.2}{79.9} \times 300 + 1{,}700 = 1{,}858 \text{ psia}$$

This confirms the assumed $P_k = 2{,}000$ closely enough for this illustration. The more closely the operating pressure approaches convergence pressure, the more accurately the calculated P_k must check with the assumed P_k, because in this region the K values are most sensitive to the P_k value used.

2 Natural gas production

Oil and gas are known as fossil fuels, and the most prevalent theory of their origin is organic—that is, that they originated a million years ago and earlier from living organisms in ancient seas. Falling to the seafloor as they died, these organisms built accumulations that eventually became overlain by other sediment and, because of the pressure exerted by both water and sediment, the organic material gradually became transformed into petroleum and natural gas.

Oil and gas accumulations

For oil and gas to accumulate, there must first be a source for their creation; second, a porous bed must exist, one which is permeable enough to permit the oil and gas to flow through it—the reservoir rock; and finally, there must be a trap, which is a barrier to fluid flow so that the petroleum and natural gas may accumulate.

Migration

It is generally accepted that any present accumulation of oil and gas is a result of migration of widely dispersed and relatively small individual quantities of hydrocarbons to a more concentrated deposit, such as is found in a reservoir. In some cases, the source material may be in close proximity to the present pool. However, it is believed that in most instances the organic source material from which petroleum and natural gas are formed is widely disseminated in the sediments and that accumulations are the result of the combination of small portions from near and far.

Several natural forces and conditions that assist such migration include (1) compaction of source beds by the weight of the overlying rock, providing a driving force tending to expel fluids through pore channels or fractures to regions of lower pressure and normally shallower depth; (2) gravitational separation of gas, oil, and water in porous rocks that are usually water saturated; (3) pressure differential from any cause between two interconnected points in a permeable medium; and (4) faulting of the earth's strata.

Reservoirs

Accumulation of oil and gas into a commercial deposit requires (1) a reservoir to contain the oil and gas, along with some water, and (2) a trap, which retains the oil and gas in the reservoir until discovery.

A petroleum reservoir is a rock capable of containing oil, gas, and water. To be commercially productive, it must have sufficient thickness, areal extent, and pore space to contain an appreciable volume of hydrocarbons, and it must yield the contained fluids at a satisfactory rate when it is penetrated by a well.

Sandstones and carbonates are the most common reservoir rocks. In order to contain fluids, they must have porosity. The porosity may be classified as (1) primary, that which is present in the original deposition, or (2) secondary, that which results from later physical or chemical changes, such as dolomitization, solution channels, or fracturing.

Porosity is expressed as the ratio of void space to the bulk volume of the rock, usually in percentage. Depending on the method of determination, porosity may represent either total or effective porosity. In many porous rocks, a number of blind or unconnected pores exists. *Effective porosity* refers specifically to those pores that are connected, permitting fluid passage.

Permeability is a quantitative measure of the ease with which a porous rock will permit the passage of fluids through it under a given pressure gradient. Like porosity, it is dependent upon rock grain shape, angularity, and size distribution. In addition, it is strongly dependent on the size of the grains. The smaller the grains, the larger will be the surface area exposed to the flowing fluid. The additional drag, or frictional resistance, of the larger surface area lowers the flow rate at a given pressure differential, and thus the smaller grain size results in a lower permeability.

Traps

A trap is a set of geologic conditions that has stopped the migration of oil and gas and caused them to be retained in a porous reservoir. Such traps are commonly related to structural highs (anticlines and domes) or to faults or unconformities. They may be placed in two general classes: (1) those in which the reservoir has an arched upper surface, and (2) those in which there is an updip termination of the reservoir.

A simple form of trap shown in figure 2.1 shows a vertical cross section of a porous and permeable reservoir rock (such as sandstone) that is overlain by a dense and impermeable bed (such as shale). The oil and gas probably originated at a point located downdip to the right or left of the fold. As they moved upward through pore passages of the water-filled reservoir rock, they encountered the sealing bed of shale overlying the reservoir rock and continued to move upward and laterally below the sealing surface until stopped by the attic of the fold. The gas, since it is the lightest of the three fluids, accumulates at the crest; the oil, next in density, forms a layer below the gas and above the water.

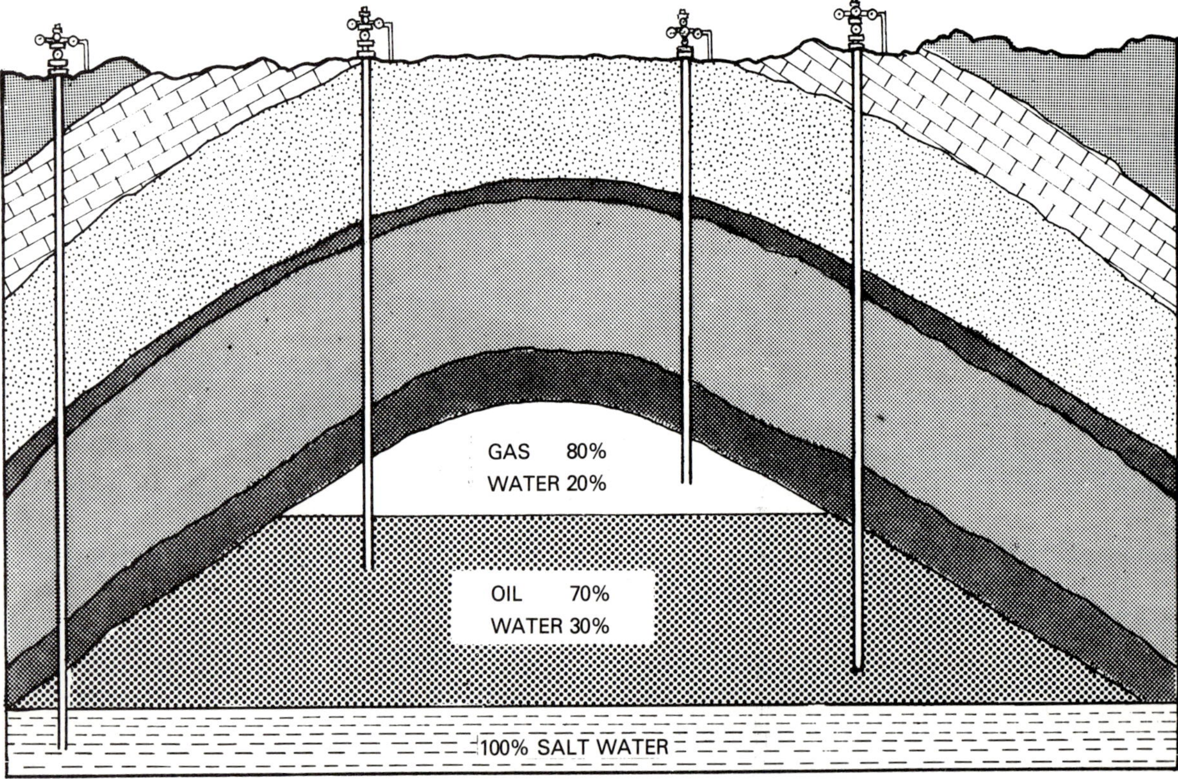

Figure 2.1. Simple oil and gas trap. Percentage figures relate to the portion of the void space in the rocks occupied by fluids.

Actually, all of the water originally contained in the pores of the reservoir rock is not totally displaced by the accumulated oil and gas. This water that remains, called *interstitial,* or *connate, water,* is of significance in making volumetric estimates of oil and gas reserves. Normally, in an intergranular type of reservoir, the interstitial water content will be in the range of 20 to 50 percent of pore space, and in some cases, either above or below this range. The connate water content is greatly influenced by the surface character of the sand grains, by the permeability of the sand, and by proximity to the level of 100 percent water.

In general, oil-wet sands have a much lower connate water saturation than water-wet sands. Also, sands of low permeability show higher connate water saturations. Furthermore, in a reservoir with very small pores, there is a transition zone at the bottom where over a vertical interval the connate water gradually increases to 100 percent. This phenomenon explains why significant percentages of oil saturation may be observed from cuttings and cores on the edge of many reservoirs, yet the production will be all water. Transition zones are more extensive and of more significance in oil reservoirs than in nonassociated gas reservoirs.

Associated and nonassociated gas

Natural gas produced from a reservoir that contains oil is called *associated gas.* The term applies to both free gas from a gas cap and solution gas. In general, the term *casinghead gas* is synonymous with associated gas, since it usually refers to gas produced from oilwells. Gas produced from a reservoir that does not contain oil is referred to as *nonassociated gas,* because it is not directly associated with oil underground. In certain fields, these terms assume particular importance because regulatory and control measures applied to associated gas are directed toward a consideration of the effect of gas production upon oil production rates and ultimate recovery, whereas the production of nonassociated gas does not involve such considerations.

Reservoir drives

The two general types of reservoir drives are *depletion drive* and *water drive.* Depletion drive is operative when the reservoir can be termed a closed reservoir—that is, the hydrocarbon accumulation is not in contact with a large body of permeable water-bearing sand. Expansion of the hydrocarbons and other reservoir materials occurring as pressure is reduced furnishes the only energy for movement of fluids through the formation and to the surface. Oil reservoirs may have either of two types of depletion drive: solution-gas drive or gas-cap drive.

Figure 2.2 represents a water drive reservoir. With the oil zone as shown, this would

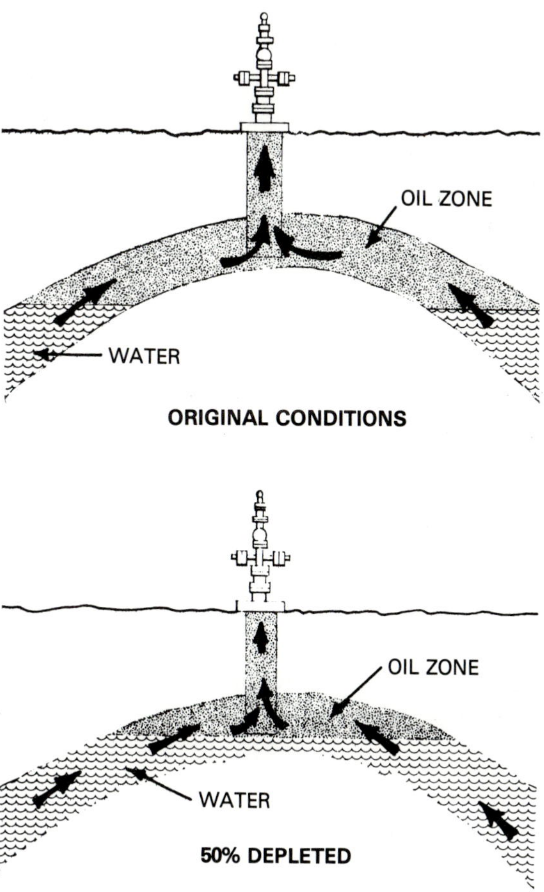

Figure 2.2. Water drive reservoir

be an oil reservoir without a gas cap. Similarly, it could be a gas reservoir with gas instead of oil accumulated in the crest of the structure. In either case, the water is free to move into the reservoir and displace the hydrocarbons as they are withdrawn. Thus, the reservoir pressure remains essentially constant, and the volume occupied by remaining recoverable hydrocarbons becomes less and less as production proceeds. Because of the greater displacement efficiency of water as compared to gas, the oil recovery from a water drive reservoir is usually much higher than for either of the depletion drives. Recovery from a good active water drive may be as much as 75 percent of the oil in place, depending on rock characteristics.

Producing gas wells

So far in this discussion, most of the features relating to origin, migration, accumulation, reservoir rock and fluid characteristics, and types of reservoir drives have been treated as though they apply to both gas and oil. Because gas and oil are so closely interrelated in nature and because a significant part of the total natural gas production is associated gas produced with oil, natural gas production is influenced in numerous ways by oil development and production practices. However, certain additional features that relate more closely to gas production should be presented.

Reservoir pressure is a controlling factor in the ability of a gas well to produce. A decrease in the bottomhole pressure of a gas well is reflected by a drop in its productivity. Reservoir pressure is always a consideration in the final stages of depletion of a gas reservoir. For most, the estimated reserves, the forecast of producing rates, and the estimated producing life are based on a prediction of pressure decline and estimated abandonment pressure. Also, considerations usually include a determination of the economic feasibility of gas compression at the wellhead to boost the pressure of the gas for sale into a pipeline and lower the abandonment reservoir pressure.

For a given production rate, pressure will decline more rapidly in a small reservoir than in a large one, if both are of the depletion type. Water influx into a gas reservoir can help support reservoir pressure and may tend to mask the difference between a small and a large reservoir, since either may have an active water drive sufficient to support reservoir pressure at the prevailing rate of production. It is possible to misinterpret a low rate of pressure decline as being due to a high gas reserve when actually the reservoir pressure is being at least partially maintained by the influx of water. The water influx causes a gradual reduction in reservoir size and a remaining gas reserve much less than that estimated on the premise of a depletion drive. Thus, any other clue that will better support a conclusion regarding the type of drive will be helpful in reaching a more reliable reserve estimate. In a multiwell gas field, a possible clue regarding the type of drive may be obtained by a study of the water production trends on different wells having various elevational levels of completion in the reservoir. An extremely rapid bottomhole pressure drop may be caused by low sand permeability, either throughout the reservoir or immediately surrounding the wellbore.

Estimation of reserves

In order to understand the art and science of reserve estimation, it is necessary to define certain terms that are commonly used and recognize the sources of the data represented by them. Some of the terms have already been discussed. Others, referring particularly to reservoir factors, are as follows:

Abandonment pressure is the average reservoir pressure at which an amount of gas insufficient to permit continued economic operation of a producing gas well is expelled. The value will vary from a few pounds per square inch gauge (psig) to 500 psig or even

1,000 psig, depending on purchaser line pressure and the amount of compression that can economically be installed.

Condensate liquids are hydrocarbons that are gaseous in the reservoir but will separate out in liquid form at the pressures and temperatures at which separators normally operate; sometimes they are called distillate. The gravity of a condensate, or distillate, is high—usually above 50°API.

Condensate ratio is the ratio of the volume of liquid produced to the volume of residue gas produced and is usually expressed in barrels per million cubic feet (bbl/MMcf).

Gas saturation is the percentage of the total pore space occupied by gas. Within the pore structure of the reservoir rock, a portion of the pore space may be occupied by gas.

Hydrocarbon pore volume is the volume of the pore space in the reservoir occupied by oil, natural gas, or other hydrocarbons (including nonhydrocarbon impurities) and may be expressed in acre-feet, barrels, or cubic feet as appropriate. An acre-foot is the volume equivalent to an area of 1 acre, 1 foot thick.

Pore volume is the volume within the reservoir (in acre-feet, barrels, or cubic feet) not occupied by rock.

Porosity is the percentage of the total reservoir that is not occupied by rock. Various methods to determine porosity, including electric logs and core analysis, are available.

Permeability is the term used to describe the flow capacity of a reservoir rock. Permeability may be reported in darcys or millidarcys. Well log correlations, core analysis, and pressure buildup (or drawdown) tests are the usual sources of permeability data. *Absolute permeability* is the permeability of a rock to a single fluid if the rock is 100 percent saturated with that fluid. *Effective permeability* is the permeability of a rock to a fluid when the saturation of that fluid is less than 100 percent.

Pressure depletion is the method of production of a gas reservoir that is not associated with a water drive. It is the process in which gas is removed and reservoir pressure declines until all the recoverable gas has been expelled.

Reservoir pressure is the average pressure within the reservoir at any given time. Determination of this value is best made by bottomhole pressure measurements with adequate shut-in time. If a shut-in period long enough for the reservoir pressure to stabilize is impractical, then various techniques of analysis by pressure buildup or drawdown tests are available to determine static reservoir pressure.

Reservoir temperature is the average temperature within the reservoir and is measured during logging, drill stem testing, or bottomhole pressure testing using a bottomhole temperature recorder.

Retrograde reservoir is a reservoir in which the pressure is high and the hydrocarbon content is completely in a gaseous or supercritical phase at initial conditions, and as the pressure drops because of production, the heavier hydrocarbon components condense, forming liquids within the reservoir. Such action is the retrograde phenomenon. If the reservoir pressure is completely depleted, only a small portion of these liquids will revaporize and be recovered.

Residual gas saturation is the portion of hydrocarbons that cannot be removed by ordinary producing mechanisms when a porous reservoir has been saturated with hydrocarbons. This value is usually a specific percentage of the pore volume. In the case of gas, the volume, measured at standard conditions, that is retained in a reservoir as residual gas saturation is an inverse function of the pressure, due to the effect of the gas laws.

Solution gas is gas dissolved in oil. Except for some very heavy oil or tar reservoirs, all oil reservoirs contain some solution gas. Solution gas may be compared to the carbon dioxide in a soft drink. If more gas is present than can be dissolved at reservoir conditions, a free gas phase will exist, and a gas cap may form. Under these conditions, the oil reservoir is said to be saturated. If less gas is present than can be held in solution, the reservoir is referred to as undersaturated.

Solution-gas drive is the mechanism in which the solution gas is the principal source of energy to expel oil from the reservoir.

Water saturation (S_w) is the percentage of total pore space occupied by water. Virtually all reservoirs contain some water. In some cases it may be the irreducible minimum, or connate water saturation. In others, a mobile water phase may be present. In all cases, water by its presence reduces the space available for hydrocarbons. Water saturation is normally estimated by electric log calculations, although cores specially obtained may be analyzed for this value.

The initial reserve estimate made on a newly discovered gas reservoir is usually a volumetric calculation. Factors needed to arrive at this estimate include type of reservoir, its area, its thickness, its porosity, and its water saturation, and the temperature, pressure, and composition of the gas. While in practice several of the following steps may be combined into a single calculation, each step here will be separately computed for purposes of illustration.

1. Determine the hydrocarbon pore volume per acre-foot of reservoir.

   $$43{,}560 \times \text{porosity} \times \frac{100 - \%\ \text{water saturation}}{100}$$

   $= \text{cubic feet/acre-foot of pore volume.}$

2. Determine the volume of gas at standard conditions initially contained in an acre-foot of reservoir.

   Gas volume (cu ft) = hydrocarbon pore space per acre-foot

   $$\times \frac{\text{original reservoir pressure (psia)}}{14.7\ \text{(psia)}} \times \frac{520°\text{R}}{\text{reservoir temperature (°R)}} \times \frac{\text{Z factor (14.7 psia)}}{\text{Z factor (original reservoir pressure)}}\,.$$

3. Determine the volume of gas at standard conditions contained in an acre-foot of the reservoir at estimated abandonment pressure.

   Gas volume (cu ft) = hydrocarbon pore space per acre-foot

   $$\times \frac{\text{abandonment pressure (psia)}}{14.7\ \text{(psia)}} \times \frac{520°\text{R}}{\text{reservoir temperature (°R)}} \times \frac{\text{Z (14.7 psia)}}{\text{Z reservoir abandonment pressure (psia).}}$$

4. Subtract volume (3) from volume (2) to obtain recoverable gas reserves per acre-foot of reservoir.

5. If it is known that an active water drive is present, the ultimate recovery is usually estimated by applying a recovery factor to the original volume in place. The recovery factor is determined by laboratory analyses on cores taken from the reservoir rock.

After production begins, it is necessary to observe the performance of the reservoir as it is depleted to determine whether the initial estimate is reasonable or should be adjusted. The typical method of evaluating performance is to plot P/Z versus cumulative recovery, where P/Z is the bottomhole pressure over the appropriate deviation factor at the time for the pressure involved. Once sufficient history has been obtained under reasonably stabilized operating conditions, it is possible to extrapolate the historical plot to the anticipated abandonment pressure and thus arrive at an estimate of reserves.

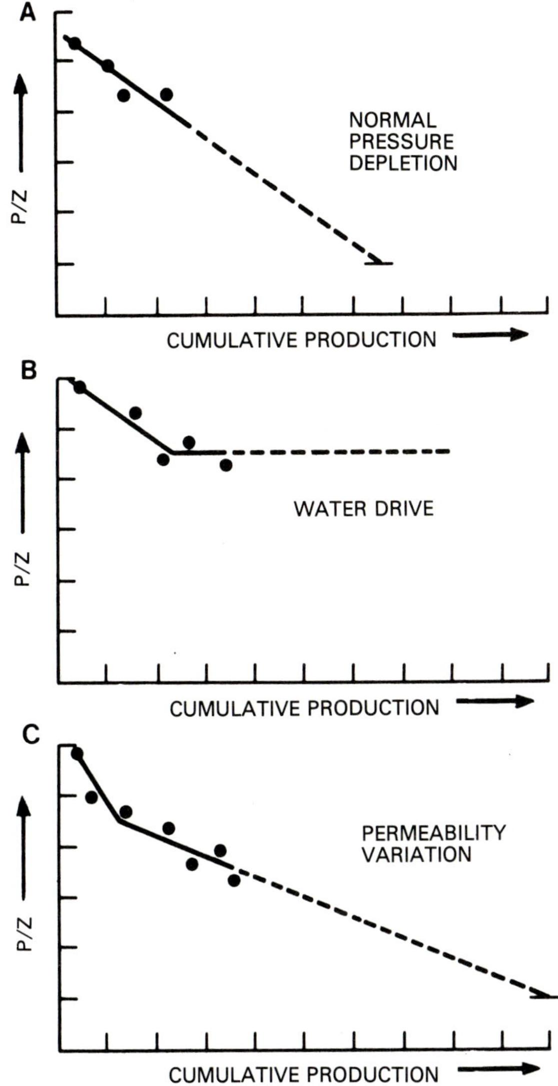

Figure 2.3. Effect of reservoir characteristics on production performance

Unfortunately, several factors will affect the validity of this method of estimation. If a full or partial water drive is present, the rate of pressure decline will be less than it would be on a straight pressure depletion (fig. 2.3). Such a decline would be erroneously interpreted as indicating a much larger reservoir than actually exists. The performance observed must thus be closely tied to the known geological configuration of the reservoir so that the reasons for the particular performance observed can be properly evaluated.

In certain other cases, it may be found that the gas reservoir is actually a gas cap on the top of an oil reservoir. In this case, the pressure decline will be influenced by the rate of oil production and the evolution of solution gas from the oil into the gas cap as the pressure declines. In such a case, the entire reservoir must be analyzed, including the effects of the oil production rate and the effect of pressure reduction on solution gas. Transfer between the oil and the gas cap can go either way, depending on reservoir conditions of temperature, pressure, and fluid composition.

Variation in the reservoir permeability can also affect the observed pressure performance. A typical pattern is a sharp drop of the reservoir pressure during the early depletion, yielding a very low ultimate recovery if extrapolated in a straight-line manner, but breaking to a lower slope on a balance between the flow capacity of the unfractured matrix and the fracture system that has been established (fig. 2.3*C*). In such cases, it is necessary to pass this breakpoint before any kind of reliable pressure decline estimate can be made.

In order to determine reserves by the pressure decline versus cumulative method, one should have accurate reservoir pressure data based on shut-in periods that are sufficient for the wells to achieve static conditions and accurate cumulative production histories for all wells in the reservoir. Although current practices normally yield adequate production data, bottomhole pressure data is often sparse and even when available is based on short shut-in periods. The most usual test provided is a shut-in wellhead pressure test obtained to meet regulatory requirements or as a result of shutting in the well for some other purpose.

In the event that the well produces a dry gas without condensate or water, it is a very simple matter to arrive at a reasonable estimate of bottomhole pressure from a shut-in wellhead pressure value. If the well is producing condensate and/or water with the gas stream when it is shut in, liquids will usually accumulate in the bottom portion of the wellbore, accounting for a substantial part of the bottomhole pressure. Unless the volume of the liquid is known reasonably well, it is impossible to estimate the bottomhole pressure from the shut-in surface pressure. In such situations, bottomhole pressure recordings are imperative.

Gas well rating

The performance of a natural gas well depends upon the physical properties of the reservoir rock, the extent and geometry of the drainage area, the properties of the flowing fluid, and the conditions of pressure distribution within the drainage matrix. The relation between the daily rate of delivery of the gas and the pressure drop within the reservoir is characteristic of the behavior of each well and is usually referred to as the back-pressure behavior of the well. A test made to determine this relationship quantitatively for a given well is referred to as a *back-pressure test.*

A back-pressure test consists of determining the static formation pressure (P_f) when the well is shut in and the flowing bottomhole pressure (P_s) at each of two or more flow rates (Q). (Note: The s in P_s is derived from "sand face.") A value of $P_f^2 - P_s^2$ is then calculated for each flow rate and plotted on log-log paper against the corresponding value of Q (fig. 2.4). It has been well established that for normal, single-phase gas wells this plotted relationship will be a straight line, the general equation of which can be written

$$Q = C(P_f^2 - P_s^2)^n$$

where C is the performance coefficient, and n is the exponent corresponding to the slope of the straight line plotted on log-log paper.

The determination of this relationship for a given well permits the analysis of many operating problems and also provides a basis for predicting future behavior of a well or group of wells. For example, once correct

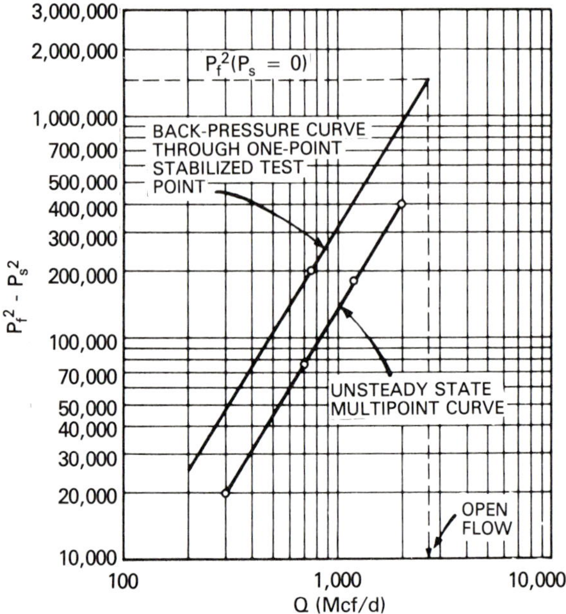

Figure 2.4. Isochronal back-pressure test for gas wells

stabilized values of C and n are established for a given well, the equation may be used as a basis for calculating (1) the effect of different sizes of producing string on the gas flow rate attainable from a well; (2) the effect of surface back-pressure on flow rate, or in other words, the gas deliverability of the well against any chosen back-pressure such as gas line pressure at the surface; (3) flow rate or deliverability of a well to be expected at some future time when the static pressure has declined to some chosen value; and (4) maximum flow rate, normally termed *absolute open flow,* that a well could theoretically deliver with zero pressure at the sand face. Although a back-pressure test of this type is relatively simple and the resultant C and n values are usable directly for a well that stabilizes quickly, the test has been found to require special treatment when dealing with tight wells—that is, wells that produce from reservoirs of low permeability and thus stabilize slowly. Such wells show slow buildup of bottomhole pressure on a buildup test, and the flowing pressure does not stabilize quickly at a constant value on a flow test.

For slow-stabilization wells, a method called *isochronal testing* has been developed whereby a true value of *n* can be determined. With this method, the well is opened at one restricted rate of flow, and pressure data are taken at specific time intervals without disturbing the rate of flow. The well is then shut in and allowed to return to a shut-in condition comparable to that existing at the time the well was first opened. It is then opened at a different rate of flow, with data being taken at the same time intervals as before. This procedure is repeated as many times as necessary to obtain the desired number of data points.

Extensive testing has proved that points taken at comparable time periods at different flow rates will yield a true value of *n*, whereas true values of *n* would not be determined if the points were taken randomly with regard to time, as on a conventional back-pressure test. A constant *n* value was obtained by using only the points read at one time when the various flow rates were used. From a practical standpoint, except for check purposes, more than one time reading at each flow rate would not be necessary in order to determine *n*.

In general, the performance coefficient *C* is a function of (1) effective permeability; (2) formation thickness; (3) gas viscosity; (4) gas gravity; (5) gas compressibility factor; (6) temperature of formation; (7) radius of drainage; (8) wellbore radius; and (9) a factor *v*, which varies with flow rate *Q* and slope *n*.

The factors *C* and *n* are considered to be constant throughout the life of a well, and essentially they are, except in certain situations that sometimes happen on any producing well. For example, gradual plugging of porosity near a well by scale or wax deposits may materially change well flow characteristics. Also, condensation of water or liquid hydrocarbons near a wellbore may have a similar effect. Certain stimulation treatments such as fracturing or acidizing may often change the flow characteristics of a well, and a retest is required to determine applicable *C* and *n* values.

Open-flow and deliverability tests of gas wells are often required by regulatory agencies, and specifications for such tests are usually stipulated. Gas sales contracts may provide for determination of gas deliverability of each connected well. Allowables assigned by regulatory bodies and rates of gas taken by gas purchasers are often apportioned on the basis of open-flow capacity or deliverability as determined by these tests.

Accuracy of a test may be affected by several factors, which include—

1. skill and experience of the tester;
2. proper testing equipment;
3. wellbore conditions, particularly presence or absence of liquids;
4. accuracy of pressure determinations; and
5. high content of nonhydrocarbon material (H_2S, N_2, and CO_2) in the gas.

Well equipment

Gas well producing problems literally start at the bottom of the well. The selection and size of casing, tubing, and wellhead equipment must take into account the expected rates of flow, fluid erosion, and chemical corrosion. Casing size will limit the size of tubing that can be installed, and tubing size will limit the gas flow that can be produced with a given pressure drawdown. Safety shutoff equipment, some of which functions by flow velocity, is mandatory by government regulations in many cases and desirable in most gas wells. This equipment is usually installed in the tubing and, in many cases, at the wellhead. High rates of flow can cause problems of erosion in the tubing and at the wellhead. Corrosion due to chemical action can be serious in gas wells, and its effects must be taken into account at all points in the flow system. Deep wells usually have more serious corrosion problems than shallow wells because the flowing temperatures are higher. Deep well gas sometimes contains organic acids, hydrogen sulfide, and carbon dioxide that will cause serious corrosion problems if protective measures are not taken. Some of the means to mitigate gas well corrosion are special alloy steels, plastic coatings, chemical injection, and displacement of chemicals into the well to obtain a protective coating on the tubing and other parts of the well exposed to corrosive fluids.

Tubing and packer

Gas well producing capacity depends upon reservoir characteristics, bottomhole pressure, restrictions of the tubing string, and pressure required at the wellhead for flow into a pipeline. If the producing pressure at the surface is less than pipeline pressure, compression will be needed to produce the well. Producing rates may vary widely, depending upon well capacity, the gas sales contract, and the daily take of the transmission line. Production from a low-pressure shallow well will usually not be very variable because of the narrow margin of available pressure. A deep gas well with high formation pressure may be required to produce 500 Mcf one day and 5,000 Mcf or more the next. Varying producing rates can cause wide temperature deviations, with varying stress on the tubing string because of expansion or contraction as the temperature of the flow stream changes. The tubing and packer arrangement in the well must be able to handle temperature variations without damaging the integrity of the tubing string.

Usually the final string of pipe run in a well is the tubing. Its small diameter permits more efficient production of fluid than that of casing and makes possible a safer completion because fluid may be circulated down the tubing and up the casing to remove undesirable fluid in the well. Tubing can be removed if it becomes plugged or damaged. In conjunction with a packer, it keeps well fluid and pressure away from the casing, because the packer seals the annular space between tubing and casing.

Plastic linings have been used in the past to minimize the damaging effect that corrosive compounds in the well stream have on tubing. Experience has shown, however, that it is difficult to obtain complete protection with plastic coatings; frequently the places left uncoated become hot spots where corrosion proceeds at accelerated rates. Also, wireline and tools run into the tubing string often damage the plastic coating. Some operators use uncoated API J-grade tubing and rely upon chemical inhibitors to minimize the damaging effects of acid gas (H_2S and CO_2) corrosion.

Corrosion inhibitors

Injection of chemical inhibitors to counter the chemical reaction between acid solutions in the gas and iron of the tubing is usually accomplished at the wellhead, either by batch treatments or continuous injection.

Batch treatment is done by injecting specified amounts of inhibitors down the tubing, displacing the well fluids in batches. They form a coating on the inside of the tubing for both physical protection and chemical inhibition. Squeezing the injection batches through the perforations into the producing interval permits longer periods of time between treatments than mere batch displacement, but allows formation damage and reduction of productive capacity.

Inhibitors can also be injected continuously down the annulus between casing and tubing in some cases. Sometimes a kill string (small-diameter tubing) is used inside the production tubing. However, a kill string reduces the capacity of the tubing for gas flow, and it cannot be used below 12,000 feet because at that depth the small-diameter pipe would pull apart under its own weight.

Wellhead equipment

An assembly of valves and fittings called a Christmas tree is installed at the wellhead to enable shutoff and regulate flow from the

Figure 2.5. Positive and adjustable chokes on a triple-zone completion

well. If corrosive conditions are expected, this equipment will be fitted with stainless trim to extend its service life. Figure 2.5 illustrates a high-pressure wellhead. Flanged or studded fittings have been standardized (table 2.1) by the American Petroleum Institute (API), following sizes and ratings established earlier by the American National Standards Institute (ANSI).

A *choke* is a restriction used in the Christmas tree for the purpose of regulating flow. Chokes may be positive (fixed size) or adjustable (fig. 2.5).

Flow regulators are sometimes employed as wellhead equipment so that a gas well may obtain rates of flow according to programmed requirements for a computerized flow controller. Regulators are normally motor-operated diaphragm valves.

TABLE 2.1
Size and Pressure Rating of Fittings

Designation		Pressure Rating, psi, 100°F	
ANSI	API	ANSI	API
600	2,000	1,440	2,000
900	3,000	2,160	3,000
1,500	5,000	3,600	5,000
- -	10,000	- -	10,000
- -	15,000	- -	15,000

Safety valves are frequently installed as wellhead equipment to obtain automatic shutoff in case of overpressure or underpressure conditions downstream of the well (fig. 2.6). This protection will operate when a flow line is plugged by hydrates or when it ruptures, making a blowout possible. Governmental regulations require automatic shutdown equipment on offshore wells for protection in case of fire or storm damage. Such equipment may also be used downhole in the tubing string.

Corrosion detection devices are frequently installed as wellhead equipment in order to obtain information as to the corrosion that

Figure 2.6. Wellhead safety shutdown valve

may be taking place in a well. These devices include containers to accumulate water samples on which iron-count observations can be made and various types of metal strips (coupons) to permit assessment of corrosive chemicals in the well stream.

Gathering systems

The equipment with which gas is collected from wells, controlled and conditioned as required, and transported to its primary destination is called a gathering system. Each gathering system is unique; it must be designed and installed to function reliably, safely, and economically within the environment of a particular field and set of physical and economic conditions. Gathering systems normally fall into one of four categories: (1) the single trunk system with laterals; (2) the loop system, in which the main line is in the shape of a loop around the field; (3) the multiple trunk system, in which several main lines extend from a central point; and (4) combinations of these. Selection of the most desirable layout requires an economic study of several possibilities and depends on many variables, such as the type of reservoir, the shape of the reservoir, the way in which the land over the reservoir is being used, the quantity of impurities in the gas, the available and permissible flow rate of gas and liquid, the flowing and shut-in pressures and temperatures, the climate and topography of the location, and the primary destination of the gas. Usually a separate economic study is needed to justify the installation of automated or remote control facilities.

Type of reservoir

Petroleum reservoirs vary from those with undersaturated crude oil, like the East Texas field, to those with gas but practically no recoverable liquids, like the San Juan Basin fields in New Mexico. Solution gas and gas from associated reservoirs tend to be rich with the heavier hydrocarbon components that liquefy easily and must often be separated before the gas can be handled by the gathering system. Gas from some nonassociated reservoirs is produced with a liquid-gas ratio low enough to permit the liquid to be gathered in the same pipe with the gas to a central point.

The shape of the reservoir, the configuration of the leases, and particularly the surface location of the wells and tanks served by a gathering system will have an influence on the system layout. For example, long narrow fields are usually best served by main trunk systems. Likewise, a group of leases around the outer perimeter of a field will probably require a loop system.

Surface usage

How the land over a reservoir is used has an important bearing on the type of gathering system employed. If the land is used for agriculture, the surface owner will probably insist that lines and other facilities be located so that they will interfere as little as possible with farming operations—planting, irrigation, and harvesting. In some locations, noisy equipment such as compressors, pumps, treaters, and heaters will need special consideration in regard to location and noise control. Some places will require special emphasis on safety devices. Gathering systems for reservoirs located beneath water are unique. Their design must take into account accessibility and protection of the environment, as well as special safety precautions.

Impurities in the gas

Impurities are common in natural gas. Some of these, such as sand and excessive amounts of water, must be removed immediately so that the gas can be gathered. Others, such as carbon dioxide, hydrogen sulfide, nitrogen, and small quantities of water, may sometimes be allowed to remain in the gas until it is gathered to a central location. There the impurities can be removed more economically in a single large plant rather than in small units located at each well.

Gas flow rate and quantity of liquids

The flow rate of associated gas is determined by the rate and method of producing the oil in the field. The flow rate of nonassociated gas production is limited by the deliverability of the well, governmental regulations, economic considerations, and contractual obligations. As gas flow rates increase, larger pipe and other equipment are required. If large quantities of liquid are allowed to flow with the gas in the pipe, flow becomes erratic, the liquid moves in slugs, and gas flow efficiency decreases. The situation is particularly serious in hilly country. To prevent or minimize the problem, scraping or slug-catching equipment is installed, or facilities are installed for handling the liquids separately.

Pressure and temperature

Gas produced at high pressure has more value because the potential energy expressed as pressure can be used to move the gas to another location for processing or use. Of course, stronger piping and valves are necessary to control the gas. Sometimes pressure in excess of that required to move the gas is expended by using a pressure-reducing, or throttling, valve so that lighter pipe may be used for the flow system. If so, it may be necessary to install heaters to prevent the formation of hydrates in the gas, which has become cold because of the refrigeration effect of a rapid drop in pressure.

Temperature of the flowing gas also affects the design of the system. High temperature necessitates installation of equipment to control movement of the pipe caused by expansion of the hot steel. On the other hand, low gas temperature presents the potential problem of hydrate formation in the gas and consequent flow stoppage. To minimize the hydrate problem, gathering lines without sharp bends are installed where hydrates may accumulate, and lines are often buried deep to minimize heat loss during cold weather. Other, more costly methods are the addition of heat by means of heaters, the injection of hydrate-inhibiting chemicals such as methanol or glycol, and the removal of water vapor by the use of dehydration equipment.

Climate and topography

Cold climates present continual, or at least seasonal, difficulties in relation to hydrates, equipment operation, and worker exposure. The climate must be considered in designing the gathering system. Location of the system in relation to topographical features, such as mountains, swamps, open water, rocky terrain, and remote sites, also affect system design.

Producing equipment

The gathering system facilities can be grouped in two major categories. First, there are those items of basic equipment that are needed to transport the gas; to start, stop, and measure the flow; and to control and measure gas pressure and temperature. Second, there is the equipment needed to condition the gas so that it will flow safely and reliably.

Natural gas has two primary commercial uses: (1) as gaseous fuel, and (2) as chemical feedstock. Sometimes a portion of produced gas is returned to the reservoir to increase total hydrocarbon recovery. It is often economically attractive to process gas in order to separate the lightest component, methane, for use as gaseous fuel from the heavier components, which may find more valuable use as liquid fuel or as chemical feedstock. Regardless of its ultimate use, gas will usually need to be conditioned in one or more ways so that it will flow safely and reliably from the well to a process plant or transmission pipeline. The gas may require liquid and sand removal by mechanical separation, water vapor removal by dehydration, increase of temperature by heaters, drop in pressure by chokes, or increase in pressure by compressors.

Basic equipment

The pipe through which the gas flows is the basic and most important part of the gathering system. It is made of steel, selected in accordance with proper codes and regulations, and usually welded together. Other pipelines and some small lines today are coupled together with threaded joints. The diameter of the pipe is determined by the amount of gas that is to be transported through the pipe, and the thickness of the pipe wall is determined by the pressure of the gas.

Flow of the gas is stopped or allowed to continue by the use of valves. Block valves are used to stop the flow; they are usually completely open or completely closed. Throttling valves to regulate the rate of flow are usually only partially open or closed; the inner valve is often positioned by a mechanical operating device controlled automatically by instruments or sometimes by a remote control arrangement.

Various instruments are needed to measure the condition of the gas. Some of these instruments are merely indicating devices that can be useful to the operator at the job site, such as liquid-level gauge glasses, mercury thermometers, and dial-type pressure gauges. Others record particular information, such as a flow recorder, which measures the flow rate and pressure and then records that information on a paper chart. Still others use the measured information to adjust the position of a control device, such as a temperature controller adjusting the gas burner control on a heater or a pressure controller changing the position of the inner valve of a pressure-reducing valve. Safety devices are also within the general description of instruments and controls; however, their function is to prevent conditions from exceeding safe limits rather than to reduce them for operational purposes. Various governmental regulations deal with the safety of flow systems and equipment.

Some degree of automation has been used in gas gathering systems for years, such as the control of pressure, flow rates, and temperatures. Expansion of the use of such regulating equipment to permit automatic control of whole systems is now common practice. The use of electronic communications and computers often enables field operations of a very complex nature to be directed instantly from a remote location accurately and automatically. Such systems are expensive and have unique maintenance problems. On the other hand, these systems can be operated with a very small work force and can quickly, often automatically, adjust the system to wide ranges of demand. They usually provide the capability of rapidly determining the location of operating difficulties.

Gas conditioning equipment

Equipment needed to condition the gas may be located at any one of several places in the gathering system. It may be located at the wellhead or at a more central point.

Separators are vessels that function as a wide place in the pipeline so that the flowing fluids slow down and gravity separates the vapors and solids from the liquids. Some are designed to separate different liquids such as condensate from water. Separators are called by many names—scrubbers, traps, knockouts, and drips.

Heaters are usually low-pressure vessels that contain a liquid—most often water—that is heated by a gas burner using fuel from the line. Pipe of sufficient strength is coiled and placed in the hot liquid, and the gas to be heated is passed through the coiled pipe. Gas heaters may be directly or indirectly fired.

Dehydrators prevent hydrate formation by removing water vapor from the gas. The gas is brought into contact with either a liquid or a solid desiccant, which takes water vapor from the gas. The desiccant is then regenerated for reuse by applying heat, which drives off the water picked up from the gas.

Compressors are installed to increase the pressure of the gas so that it can flow through the pipeline. Friction is developed by flow through the pipe; so if the pipeline is very long, additional compressors may have to be installed along the line. Reciprocating piston compressors are the type most often used; the smaller, high-speed units are used for small volumes, the larger, low-speed, integral units for large volumes. Centrifugal compressors are becoming more common, particularly for main transmission lines.

Natural gas that contains corrosive elements such as carbon dioxide or hydrogen sulfide probably must be conditioned so that the flow system equipment does not deteriorate and become a safety hazard or an economic loss. If small quantities are involved, the conditioning may be by the injection of chemical inhibitors into the gas stream; if the quantities are large, the most economical method may be removal of the contaminant. Removal is usually done with equipment and procedures similar to those used to dehydrate gas.

Gas plant, gasoline plant, and *gas liquids extraction plant* are all terms applied to facilities that remove some of the heavier or easily liquefied hydrocarbon component from the gas so that it can be used separately from the gas. The process is carried out only when there is justification from an economic standpoint, conservation of natural resources, or the need of the consumer. See Appendix B for a more complete description of the purpose and design of gas processing plants.

3 Natural gas and liquid separation

A natural gas well stream as produced from a reservoir is a complex mixture of hundreds of different compounds of hydrogen and carbon, all with different densities, vapor pressures, and other physical characteristics. A typical well stream is a high-velocity, turbulent, constantly expanding mixture of gases and hydrocarbon liquids, intimately mixed with water vapor, free water, solids, and other contaminants. As it flows from the hot, high-pressure reservoir, the well stream is undergoing continuous pressure and temperature reduction. Gases evolve from the liquids, water vapor condenses, and some of the well stream changes in character from liquid to bubbles, mist, and free gas. The high-velocity gas is carrying liquid droplets, and the liquid is carrying gas bubbles.

Stated simply, field processing is for the purpose of removing and separating the well stream into salable gas and petroleum liquids, recovering the maximum amounts of each at the lowest possible overall cost. Field processing of natural gas actually consists of four basic processes:

1. Separation of the gas from entrained solids and free liquids such as crude oil, hydrocarbon condensate, and water
2. Conditioning the gas to remove other undesirable components, such as hydrogen sulfide or carbon dioxide
3. Conditioning the gas to remove condensable water vapor, which under certain conditions might cause hydrate formation
4. Processing the gas to remove condensable and recoverable hydrocarbon vapors

Conventional separators

Separation of well stream gas from free liquids is by far the most common of all field processing operations and, at the same time, one of the most critical. A properly designed separator will provide a clean separation of free gases from free hydrocarbon liquids. A well stream separator must—

1. cause a primary-phase separation of the mostly liquid hydrocarbons from those that are mostly gas;
2. refine the primary separation by removing most of the entrained liquid mist from the gas;
3. further refine the separation by removing the entrained gas from the liquid; and
4. discharge the separated gas and liquid from the vessel and ensure that no reentrainment of one into the other takes place.

If these functions are to be accomplished, the basic separator design must—

1. control and dissipate the energy of the well stream as it enters the separator;
2. ensure that the gas and liquid flow rates are low enough so that gravity segregation and vapor-liquid equilibrium can occur;
3. minimize turbulence in the gas section of the separator and reduce velocity;
4. control the accumulation of froths and foams in the vessel;
5. eliminate reentrainment of the separated gas and liquid;

6. provide an outlet for gases, with suitable controls to maintain preset operating pressure;
7. provide outlets for liquids, with suitable liquid-level controls;
8. if necessary, provide cleanout ports at points where solids may accumulate;
9. provide relief for excessive pressures in case the gas or liquid outlets should be plugged; and
10. provide equipment (pressure gauges, thermometers, and liquid-level gauge-glass assemblies) to check visually for proper operation.

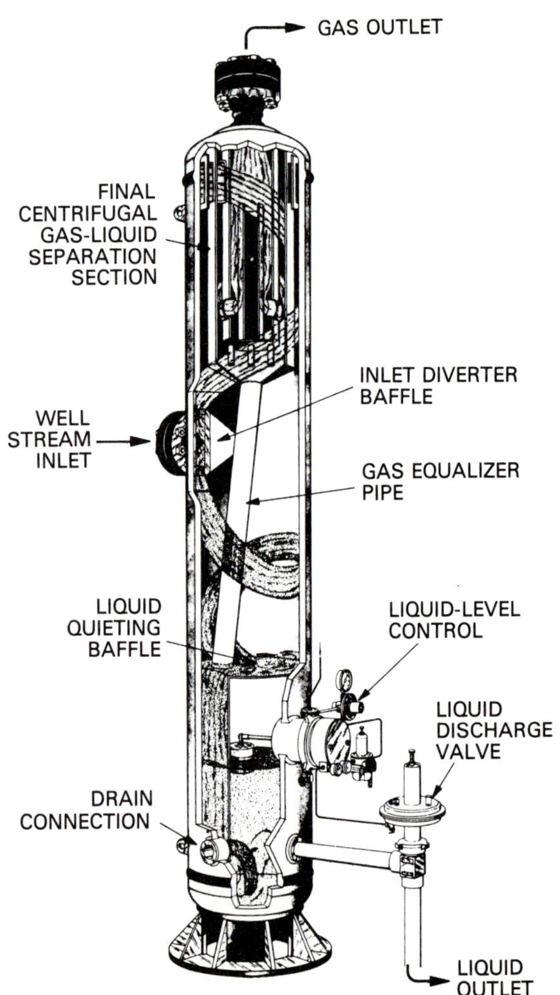

Figure 3.1. Conventional vertical separator

The process equipment and conditions downstream of a separator will usually dictate the necessary degree of separation and the actual vessel design. Ideally, gases and liquids should come to full equilibrium in the separator, but a compromise must usually be made between the degree of separation achieved and the cost of the installation.

The following factors must be considered in sizing and selecting a separator:

1. liquid flow rate (oil and water) in barrels per day, minimum and peak instantaneous readings;
2. gas flow rate in million standard cubic feet (MMscf) per day;
3. specific gravities of oil, water, and gas;
4. required retention time of fluids within the separator, retention time being a function of physical properties of the fluids;
5. temperature and pressure at which the separator will operate, and the design pressure of the vessel;
6. whether the separator is to be two-phase (e.g., liquid and gas) or three-phase (e.g., oil, water, and gas);
7. whether or not solid impurities (sand, paraffin) are present;
8. whether or not foaming tendencies exist.

Three basic types of separators are widely used for gas-liquid separation: (1) vertical (fig. 3.1); (2) horizontal (fig. 3.2); and (3) horizontal double barrel (fig. 3.3). Each has specific advantages, and selection is usually based on which one will accomplish the desired results at the lowest cost.

Vertical

A vertical separator is often used on low to intermediate gas-oil ratio well streams and also when relatively large slugs of liquid are expected. It can be fitted with a false cone bottom to handle sand production. A vertical separator occupies less floor space than others, an important consideration on an offshore platform. However, because the natural upward

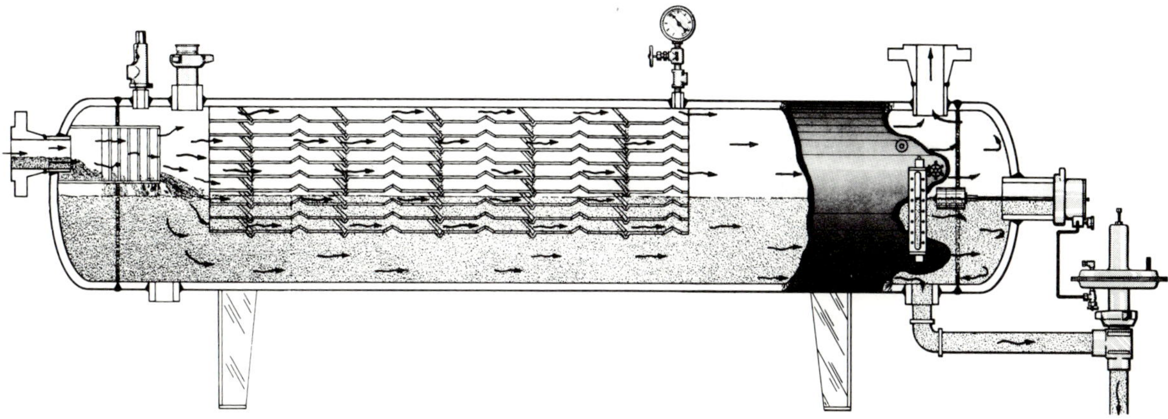

Figure 3.2. Conventional horizontal separator

flow of gas in a vertical vessel opposes the falling droplets of liquid, a vertical separator may be larger and more expensive than a horizontal unit of the same capacity.

Internal designs of vertical separators vary but a typical design can be described as follows: An inlet diverter spreads the inlet fluids against the vertical separator shell in a thin film and at the same time imparts a centrifugal motion to the fluids. These actions provide the desired momentum reduction and allow the gas to escape from the thin oil film. The gas rises to the top of the vessel, and the liquids fall to the bottom. Some small liquid particles are swept upward with the rising gas stream, and these particles are separated by a centrifugal baffle arrangement below the gas-outlet connection.

Horizontal

The horizontal separator is perhaps the best separator for the money. It may be less expensive than the vertical separator for equal capacity. It has a much greater gas-liquid interface area, consisting of a large, long, baffled gas separation section, permitting

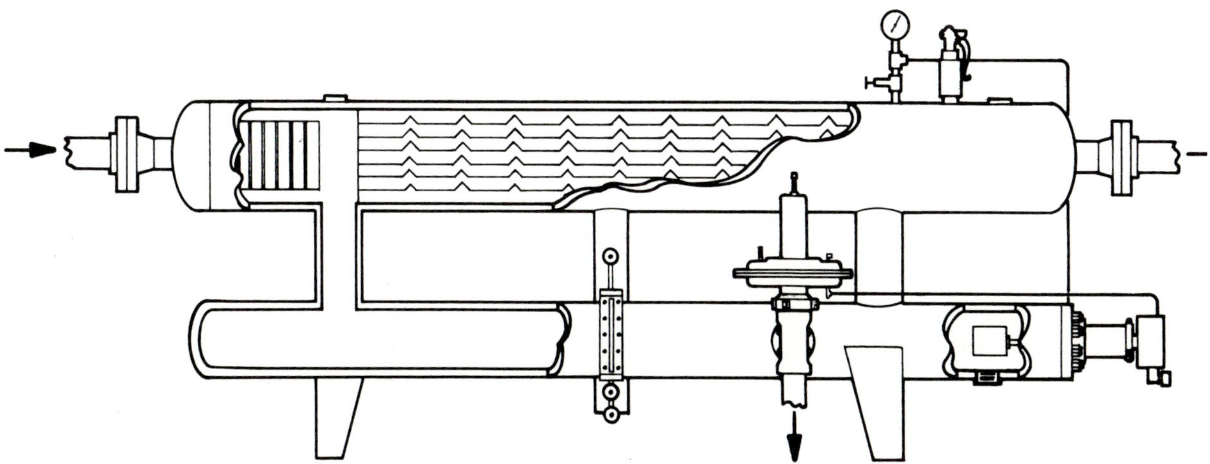

Figure 3.3. Conventional horizontal double-barrel separator

much higher gas velocities. Horizontal separators are almost always used for high gas-oil ratio well streams, for foaming well streams, or for liquid-from-liquid separation. A horizontal separator is easier to hook up, easier to service, and easier to skid-mount than a vertical separator. Several separators can be stacked easily into stage-separation assemblies, minimizing space requirements. In a horizontal separator, gas flows horizontally and, at the same time, entrained liquid falls toward the liquid surface. Some separators have closely spaced horizontal baffle plates that extend lengthwise down the vessel, upon which are evenly spaced baffle plates at a 45° angle to the horizontal. The gas flows on the baffle surfaces and forms a liquid film that is drained away to the liquid section of the separator. The baffles need only be longer than the distance of liquid trajectory travel at the design gas velocity.

A horizontal three-phase separator (fig. 3.4) is designed to separate oil, water, and gas and has two liquid outlets. Three-phase separators are commonly used for well testing and for instances in which free water readily separates from oil or condensate. They are identical to two-phase vessels except for the water compartment and an extra level-control and dump valve.

Some separators use knitted wire mesh pads, 4 to 8 inches thick, across the gas section of the separator or across the gas-outlet nozzle. Wire mesh can provide good gas-liquid separation but may be plugged by paraffin or solids in the gas stream. Wire mesh pads should not be used unless the well stream is so clean that no danger of plugging exists.

Double-barrel horizontal

A double-barrel horizontal separator has all the advantages of a normal horizontal separator plus a much higher liquid capacity. Incoming free liquid is immediately drained away from the upper section into the lower section. The upper section is filled with baffles, and gas flow is straight through and at higher velocities than in the lower section.

Separator controls

Liquid level within the separator must be maintained within reasonable limits to prevent the discharge of liquid out of the gas line or gas out of the liquid line and to assure proper functioning and flow through the separator internals.

Pressure within the separator is usually maintained within a specific pressure range by a gas back-pressure regulating valve.

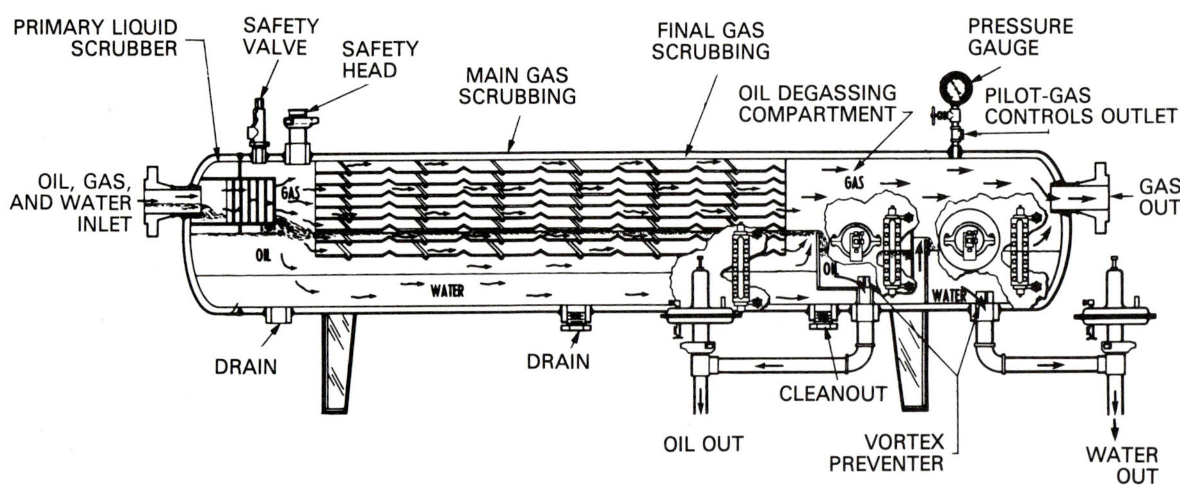

Figure 3.4. Conventional horizontal three-phase separator

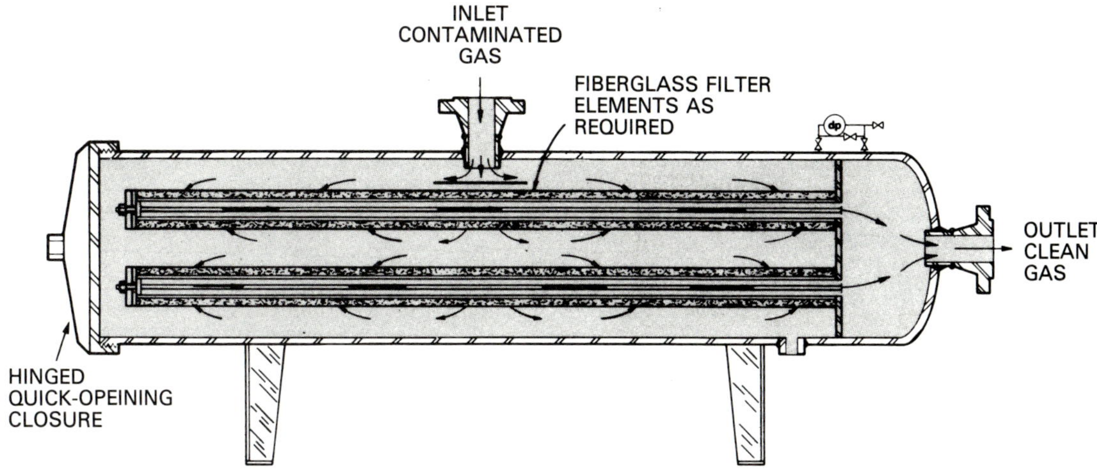

Figure 3.5. Dry-gas filter separator

Temperature within most separators is not usually controlled, although there are exceptions, such as low-temperature separation systems.

Safety and protection against overpressure is provided by a pressure relief valve set at the design pressure of the separator.

Filter separators

Filter separators are designed to remove small liquid and solid particles from gas streams. These units are designed specifically to handle those applications in which, due to the extremely small particle size, conventional separation techniques employing gravitational or centrifugal force are ineffective. Contaminants of such small size can be removed most effectively by passing the gas through a fine high-quality filtering medium. Several configurations of filter separators are used, depending upon the required efficiency and on whether liquids or solids or both are to be removed. Some filter elements have collection efficiencies of 98 percent of the 1-micron particles and 100 percent of the 5-micron particles when they are operated at rated capacity and recommended filter-change intervals.

Applications

When removal of dry solids is the only requirement, a simple dry gas filter separator is used (fig. 3.5). It has no provisions for collecting or dumping liquids. A dry gas filter separator is used in lines carrying gas undersaturated with condensable hydrocarbon and water—for example, gas from a stripping plant. It may be used where liquid is occasionally present if the passage of liquid through the filter is not objectionable. Filter elements are easily changed from the hinged quick-opening enclosure.

When both liquid and solid contaminants must be removed from gas, a two-compartment vessel is normally used (fig. 3.6). Liquid is coalesced, and solid particles are filtered from the gas by a fiberglass element. Solids are retained by the element, while liquids are agglomerated by wetting the surface of the fiberglass. Liquids are blown off the outlet end of the fiberglass element in large droplets, which are then easily captured by mist-extracting baffles. Liquids are collected in the lower barrel and dumped by a level-control and valving system. A primary liquid collection lower barrel (dotted in on fig. 3.6) is required when liquid volumes are large or when liquid enters the vessel in slugs.

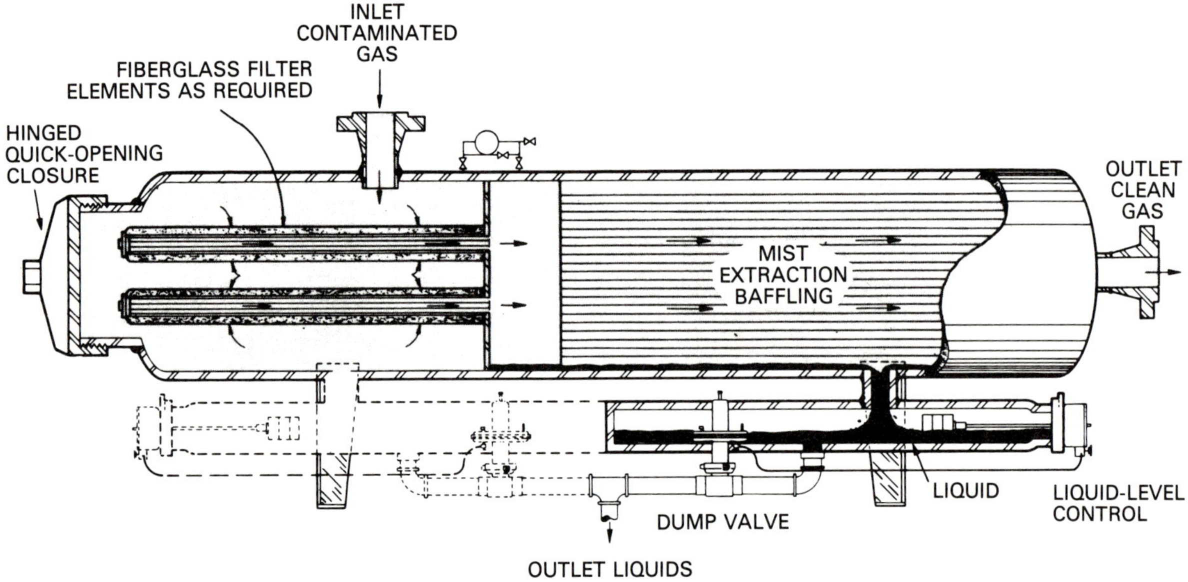

Figure 3.6. Filter separator for removing liquid and solid contaminants

Figure 3.7 illustrates a filter separator designed for fog coalescing and gas filtering. If the major problem is fog coalescing, the gas flow through the elements may be from inside out, the reverse of normal flow. In this application, the liquid collects in large droplets on the outside of the element and can fall to the bottom of the vessel without being broken up by the high gas velocity. Therefore, no mist-extraction baffling is required. Liquid flows to the catch pot in the bottom of the vessel, where discharge is controlled by a liquid-level control.

A large-capacity filter separator is shown in figure 3.8. Large capacities can often be best handled by a dual vessel arrangement. Inlets

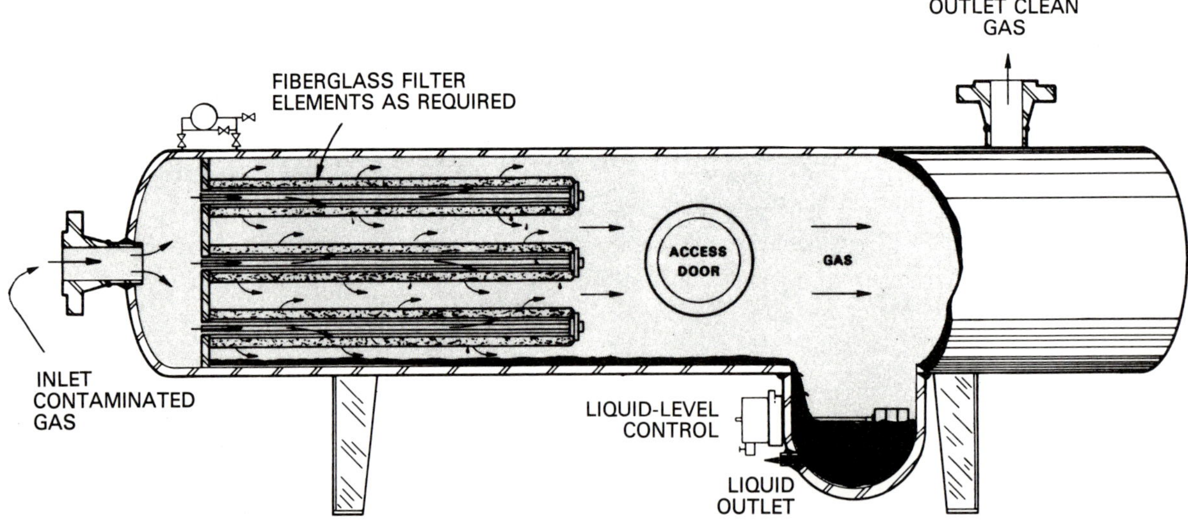

Figure 3.7. Filter separator for fog coalescing and gas filtering

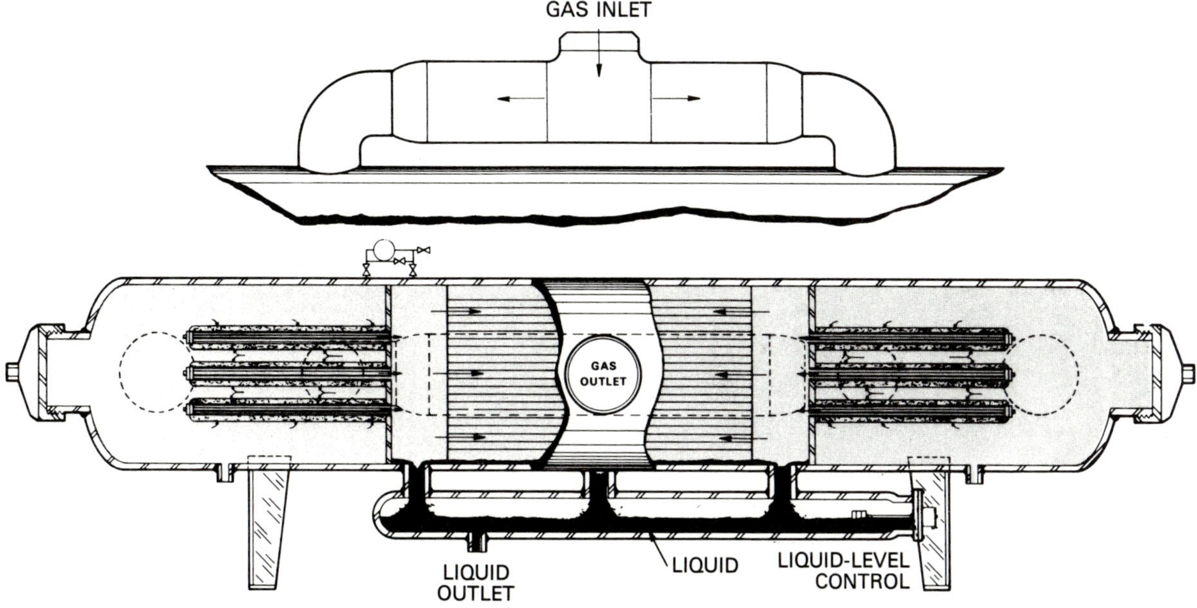

Figure 3.8. Large-capacity filter separator

are provided at each end, with flow going toward a common center outlet. Independent filter element sections are provided at each end, with quick-opening access closures for filter removal. Liquids are collected in the lower barrel and dumped by a liquid-level control and valve system.

Filter elements

A standard filter element is made of a perforated metal cylinder with gasketed ends for compression sealing. A fiberglass cylinder with a ½-inch wall is located outside the metal cylinder. The gas flow is normally from the outside of the cylinder to the center. There is a layer of fabric outside and inside the glass cylinder to prevent the migration of the glass fibers into the gas stream (fig. 3.9). When the primary duty is coalescing of liquid particles, elements with flow from the center to the outside may be used. Hoop strength is added to these elements by glass fibers wrapped around the element. Filter elements are securely held over openings in the vessel tube sheet by a center rod. This rod centers each element over its tube-sheet opening and provides the compression for sealing the element between the

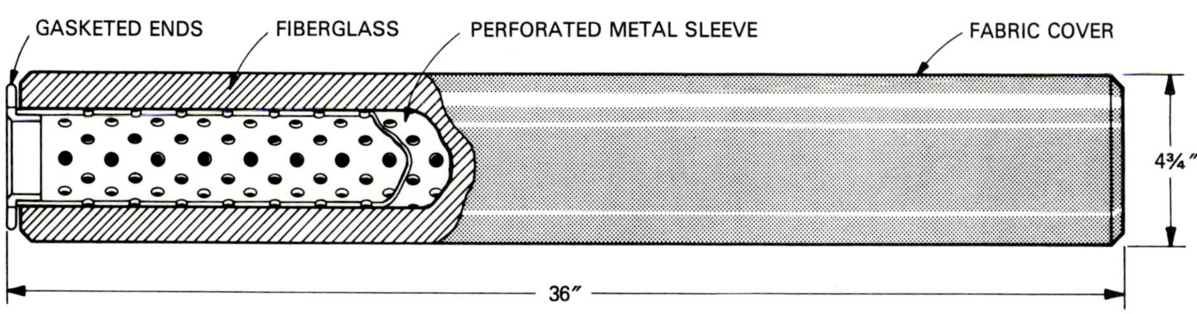

Figure 3.9. Standard filter element

TWO-STAGE SEPARATION

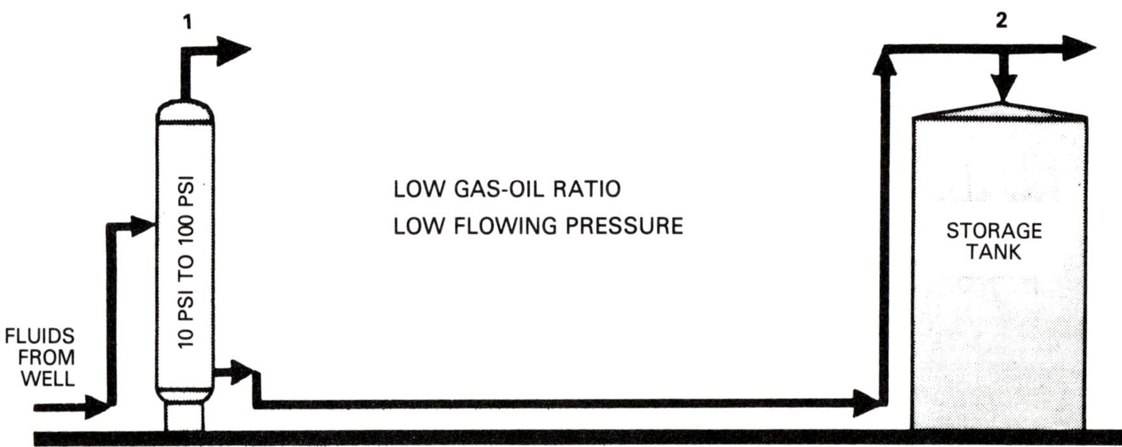

THREE-STAGE SEPARATION

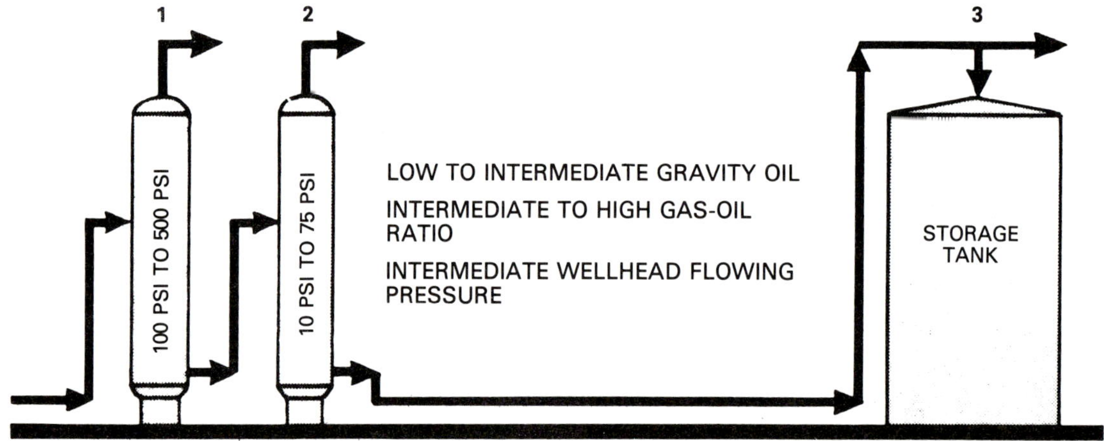

FOUR-STAGE SEPARATION

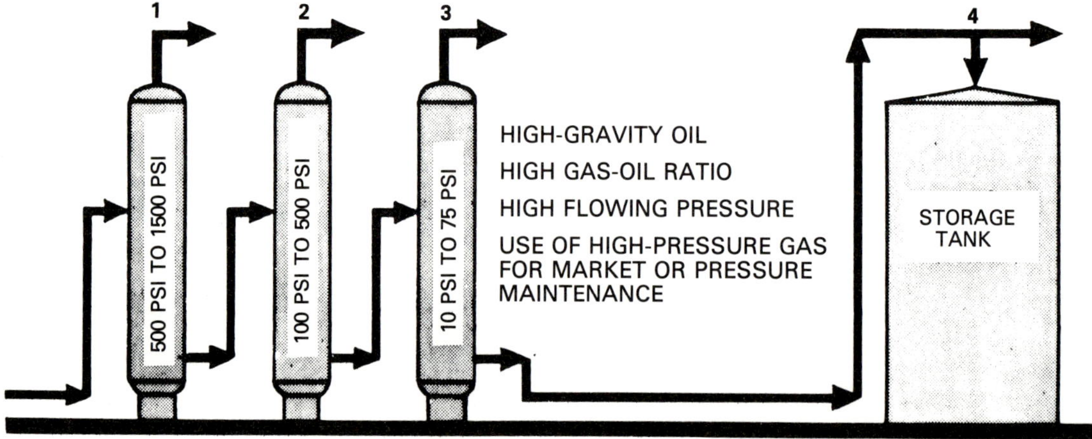

Figure 3.10. Stage separation flow diagrams

tube sheet and the plate, which closes the opposite end. The center rod is removable from the vessel for unobstructed access.

Pressure drop

Filter cartridges offer little pressure drop to the flow of gas when they are clean. Pressure drop through the entire vessel with clean elements will be 2 psi at its rated capacity unless stated otherwise. Approximately half of this pressure drop will be caused by the elements' supporting hardware. Filter elements should be changed when the differential pressure across them rises approximately 6 psi above the initial (clean) pressure drop. The differential pressure should never be permitted to reach the 30-psi element collapse pressure. This pressure can be read directly on the differential pressure gauge, which is provided. A 30-psi pressure gauge encased in a 2,500-psi working-pressure glass-front case is used for this service.

Stage separation

Stage separation is a process in which gaseous and liquid hydrocarbons are separated into vapor and liquid phases by two or more equilibrium flashes at consecutively lower pressures. Two-stage separation involves one separator and a storage tank (fig. 3.10). Three-stage separation requires two separators and a storage tank. Four-stage separation requires three separators and a storage tank. The tank is always counted as the final stage of vapor-liquid separation because the final equilibrium flash occurs in the tank.

The purpose of stage separation is to reduce the pressure on the reservoir liquids a little at a time, in steps, or stages, so that a more stable stock tank liquid will result. Petroleum liquids at high pressure usually contain large quantities of liquefied propanes, butanes, and pentanes, which will vaporize or flash as the pressure is reduced. This flashing can cause a substantial reduction in stock tank liquid recovery, depending on well stream composition, pressure, temperature, and other factors. For example, if a volatile condensate at 1,500 psig were discharged directly into an atmospheric storage tank, most of it would vaporize immediately, leaving very little liquid in the tank.

The ideal method of separation, to attain maximum liquid recovery, would be that of differential liberation of gas by means of a steady decrease in pressure from that existing in the reservoir to that existing in the storage tanks. With each tiny decrease in pressure, the gas evolved would immediately be removed from the liquid. However, to carry out this differential process would require an infinite number of separation stages, obviously an impractical solution. A close approach to differential liberation can be made by using three or more series-connected stages of separation, in each of which flash vaporization takes place. In this manner, the maximum economical amount of liquid can be recovered. Figure 3.11 shows a typical skid-mounted horizontal stage separation unit.

Low-temperature separation

Low-temperature separation, probably the most efficient means yet devised for handling high-pressure gas and condensate at the wellhead, performs the following functions:

1. separation of water and hydrocarbon liquids from the inlet well stream;
2. recovery of more liquids from the gas than can be recovered with normal-temperature separators; and
3. dehydration of gas, usually to pipeline specifications.

The first low-temperature units were developed and placed in operation in 1948 to

Figure 3.11. Skid-mounted two-stage separator unit

dehydrate gas at the wellhead in remote locations so that high-pressure gas and condensate could be gathered at central locations without the problems of hydrate plugging.

Essentially, the low-temperature separation process is one of intentional hydrate formation and controlled melting. The inlet gas is cooled by expansion due to pressure reduction, causing water and liquid hydrocarbons to condense; if hydrates are formed, they are quickly melted. Dry gas, condensate, and free water are then discharged from the vessel under controlled conditions. Since this separation system permits operating conditions well below hydrate-formation temperatures, the recovery of hydrocarbon liquids is much higher than that possible for conventional separation. Condensation of a higher percentage of the water vapor is also accomplished, resulting in dehydration of the gas.

Since low-temperature separation is achieved by the cooling effect of gas expansion—that is, pressure reduction across a choke—how much pressure drop is needed to obtain adequate cooling? Generally speaking, satisfactory operation and dehydration to pipeline specifications can be accomplished with a pressure differential as low as 1,000 psi, provided the temperature upstream of the choke can be controlled close to the hydrate point.

The actual minimum differential between flowing and line pressures depends primarily upon the pressure range in which the hydrate temperature falls. On the pressure-temperature drop chart (fig. 3.12), it can be seen that a 1,500-psi drop from 3,000 psi at 120°F will provide a final temperature of only 78°F, while a 1,500-psi drop from 2,000 psi at 120°F will give a final temperature of 49°F. The actual temperature drop per psi of pressure drop will not necessarily correspond to that shown on the chart; composition of the gas stream, flow rate, liquid rates, bath temperature, and ambient temperature will affect the actual temperature drop. Water dew points will average 10°F to 15°F below the indicated temperature because of the adsorptive effects of the hydrates on the vapor-phase water.

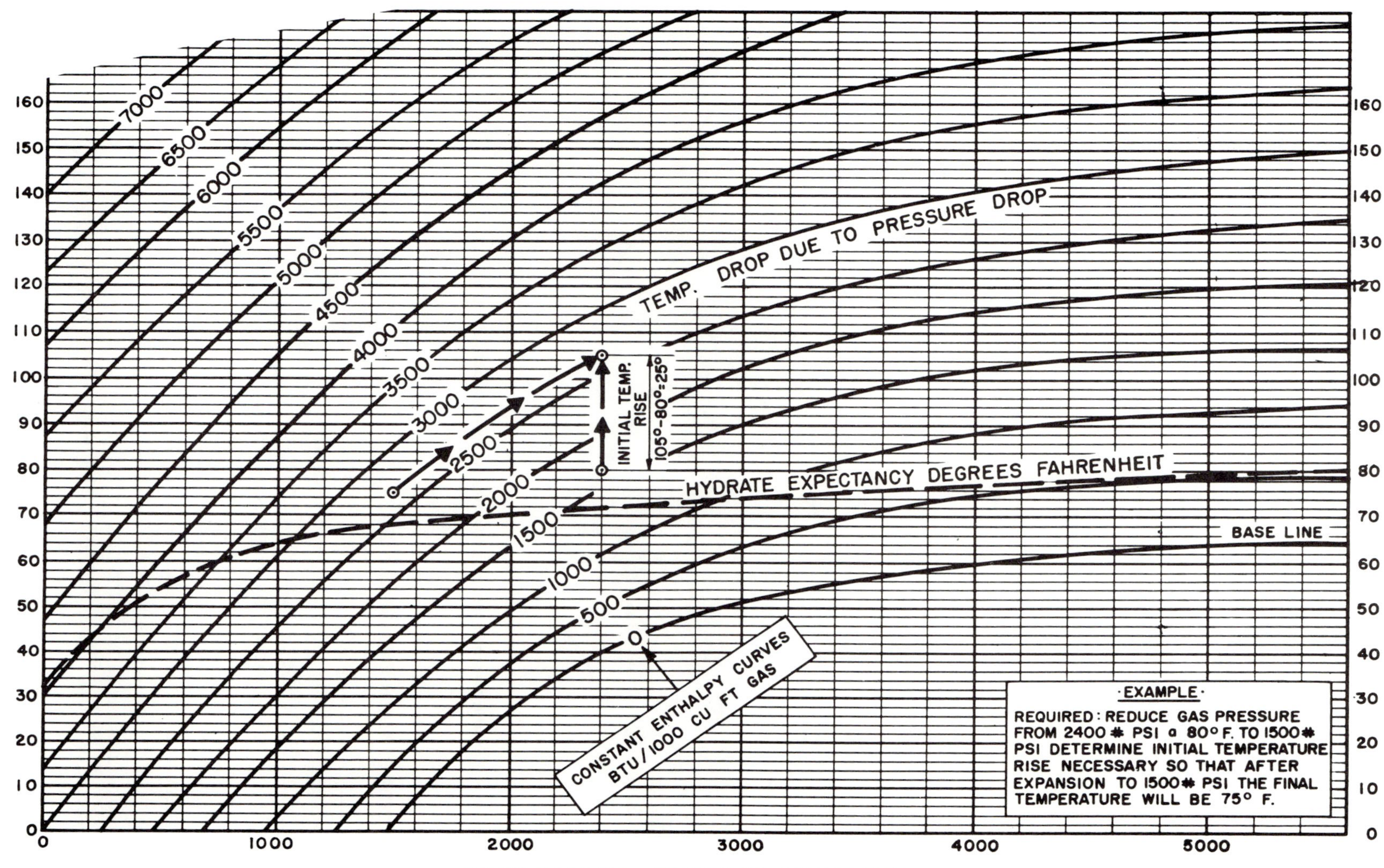

Figure 3.12. Natural gas expansion–temperature reduction curves based on 0.7 specific gravity gas

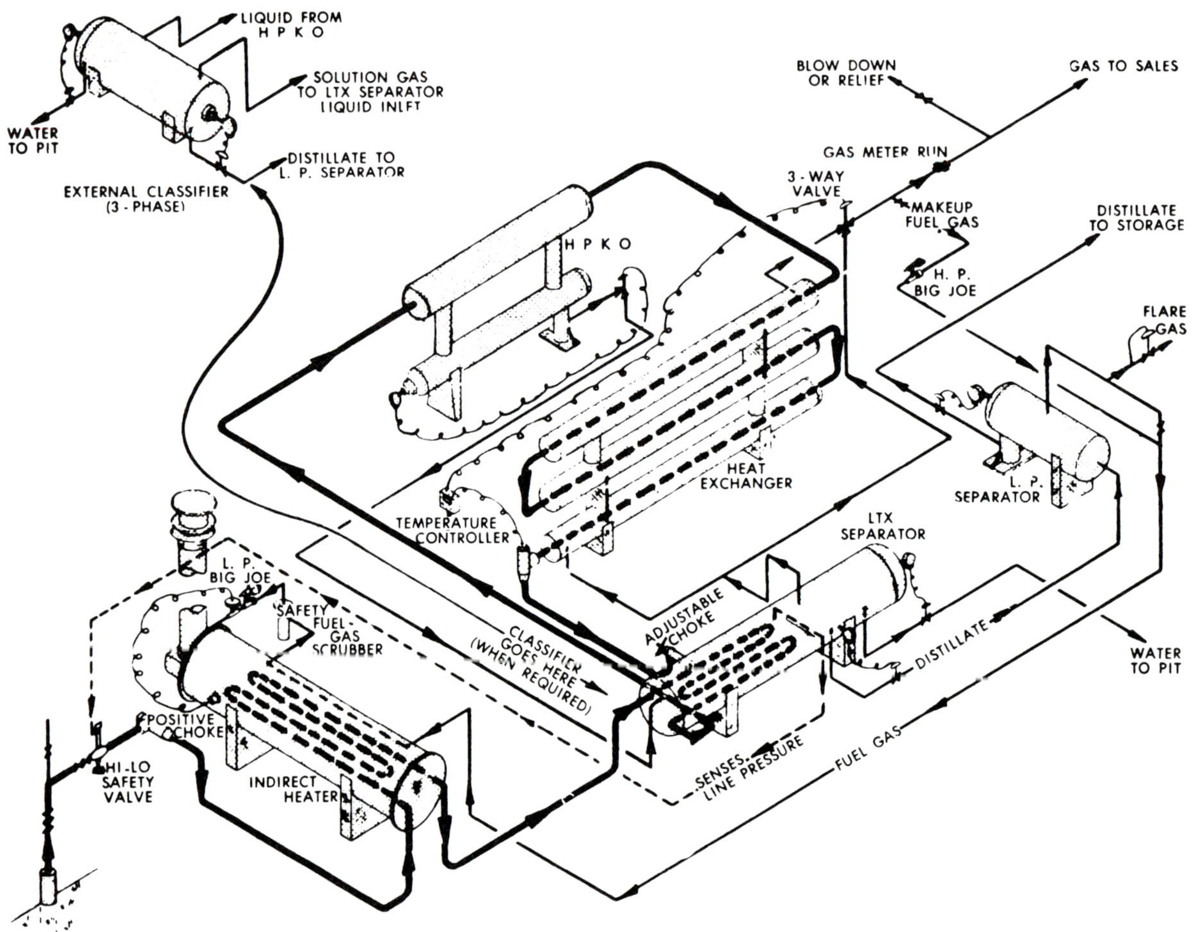

Figure 3.13. Schematic flow diagram of a low-temperature separation unit

Low-temperature systems

Low-temperature systems may be discussed in two categories: (1) those that use hydrate inhibitors and (2) those that do not. The same principle of refrigeration by expansion is utilized in both systems, but the details differ sufficiently to warrant separate discussion. A third category of low-temperature separation systems, called the mechanical refrigeration system, operates the same as the first two; the only difference is that the expansion of high-pressure gas is not used to cool the inlet well stream. Instead, a separate closed-cycle propane or ammonia refrigeration system is used to cool the well stream before separation. A mechanical refrigeration system is often used when well stream flowing pressures are too low to obtain adequate pressure differential and expansion for cooling.

A schematic flow diagram of a typical low-temperature separation system, using no hydrate inhibitor, is shown in figure 3.13. The basic equipment includes—

1. inlet well stream heater (optional);
2. inlet free liquid separator;
3. heat exchanger, which changes cold sales gas to warm inlet gas;
4. choke or other expansion valve on inlet to low-temperature separator;
5. low-temperature separator with heat-exchange coil in liquid section;

6. flash separator; and
7. piping and controls.

Depending upon the flowing temperature of the inlet well stream, the inlet well stream heater may or may not be required. The well stream must be warm enough to melt any hydrates that may form in the bottom of the low-temperature separator. Assuming that a heater is required, the inlet well stream flows through the heater coil and then through a coil located in the bottom of the low-temperature separator (fig. 3.14).

The well stream provides heat to warm the separator liquids and melt hydrates that may have formed. In passing through this coil, the well stream is partially cooled, causing condensation of some water and liquid hydrocarbons.

The well stream then flows to a high-pressure liquid knockout, where the liquids may be handled in one of three ways.

1. Water only is removed in the knockout; the condensate goes across the choke with the gas.
2. All liquids are removed in the knockout and are then dumped into the liquid section of the low-temperature separator.
3. All liquids are removed in the knockout, with the water being dumped to disposal and the condensate dumped to the liquid section of the low-temperature separator.

The well stream then leaves the knockout and flows through the tube side of a sales-gas/inlet-gas heater exchanger, where the inlet well stream is cooled by the cold gas leaving the low-temperature separator. Temperature control of the well stream is very critical at this point, since a temperature too low will result in hydrate formation and plugging of the exchanger tubes or the choke. A temperature too high will result in inadequate dehydration of the gas and reduced recovery of condensate.

After being properly cooled, the well stream flows through the choke, expanding directly into the low-temperature separator, dropping from inlet pressure to sales-line pressure, and cooling as a result of the expansion. This cooling causes some of the hydrocarbon gases and most of the water vapor to liquefy and, in most cases, causes hydrates to form. The hydrates, water, and condensate are deposited in the warm liquid bath where the hydrates are melted and the condensate and water are separated.

The cold dehydrated gas leaves the low-temperature separator, flowing through or partially bypassing the previously mentioned gas-to-gas heat exchanger. The volume of cold sales gas flowing through the shell side of

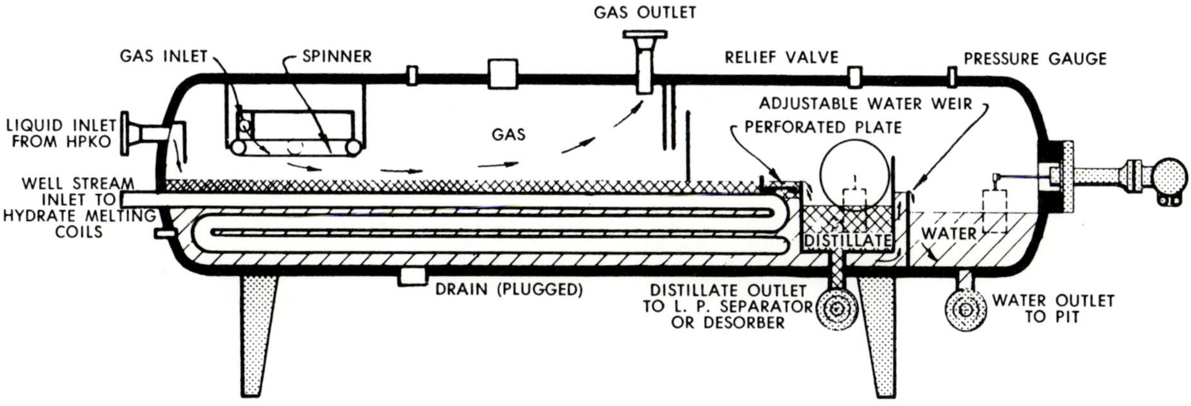

Figure 3.14. Cross section of a low-temperature extraction separator

this exchanger is determined by a three-way proportioning valve. The proportioning valve is controlled by a temperature controller whose bulb is in the well stream immediately upstream of the choke. The excess cold gas, which is not required for cooling the well stream, is diverted through the bypass piping and then recombined with the cooling gas. The gas then flows through an orifice meter to the sales line.

The hydrocarbon liquids, which were dumped from the inlet free-liquid knockout, are dumped into the liquid section of the unit. The evolved solution gas from these hydrocarbon liquids then joins the main gas stream, which has already expanded across the choke.

The relatively long residence time of the hydrocarbon liquids in the low-temperature separator, plus the heat from the warm hydrate-melting coils causes the condensate to reach a considerable degree of stabilization in the vessel. The hydrocarbon components, which may have been vaporized by the heating coil, will be recondensed by the cool gas and will remain in the liquid phase.

Water from the separator is discharged directly to disposal facilities. All controls and instruments are operated from a pilot gas system using dehydrated gas, which has been warmed in the heater.

Hydrocarbon liquids from the low-temperature separator are usually discharged to a second-stage low-pressure flash separator to conserve liquid volume. Alternately, some installations may use a condensate stabilizer if a controlled vapor pressure product is required. Condensate will then be discharged from the flash separator and stabilizer to the liquid storage tanks.

In order to conserve gas, fuel for the inlet heater and instrument supply gas is withdrawn from the flash gas stream leaving the low-pressure separator. Well streams producing 10 bbl per MMscf or more will usually provide sufficient flare gas to operate the heater and controls.

Figure 3.15. Skid-mounted low-temperature separation system

The gas volume flowing through a low-temperature unit (fig. 3.15) may be regulated by a manually controlled or pilot-operated choke. Manually controlled chokes are normally used when both volume of flow from the well and pipeline delivery pressures are constant. If delivery line pressure does not change and wellhead pressures remain constant, the rate of flow through the unit will be constant for a given choke setting. When wellhead pressures are variable or when pipeline pressures vary from day to day, it is often necessary to regulate flow through the unit by a pilot-operated choke or a pressure-reducing regulator.

Glycol injection system

The system in which a hydrate inhibitor is used provides for injection of glycol prior to the expansion of the gas through the choke. This eliminates the necessity for a heating coil in the bottom of the low-temperature separator. It is necessary in this system to inject sufficient glycol to prevent hydrate formation ahead of the choke and in the separator. In this system, the glycol is separated and sent to a regenerator where the absorbed water is driven off by heating so that the glycol can be reused.

The volume of glycol to be injected into the well stream depends upon the inlet conditions of pressure and temperature; these factors determine the amount of water present. In systems with inlet temperatures in the range of 90°F and pressures of 1,700 psi, approximately 5 gallons of 85 percent (by weight) diethylene glycol is required per 1 million cubic feet of gas. This amount increases with increasing temperature and decreases with increasing pressure. Most installations are operated with injection of 6 to 10 gallons per 1 million cubic feet.

Glycol losses in the system occur in three ways: (1) solution in condensate; (2) carryover in the gas stream; and (3) losses in the regenerator. The solubility of diethylene glycol in condensate is approximately 1 gallon per 100 barrels at 60°F and 100 psi. Carryover in the low-temperature separator and in the regenerator is dependent upon the design of the equipment to a large extent but is normally less than 0.2 gallon per 1 million cubic feet.

In some systems, glycol is injected into the gathering system at or near the wellhead, and the gathering system is operated with minimum hydrate troubles to a central point in the field for separation of condensate and gas. In this type of system, the volume of glycol required may be reduced by the installation of a free-liquid knockout prior to the glycol injection point to remove free water produced from the well.

When there is no provision for melting hydrates in the low-temperature separator of a glycol injection system, some operators use a hydrate detector between the choke and the low-temperature separator to prevent plugging the separator. A hydrate detector is simply an enlargement in the pipe, filled with baffles or ceramic rings. In the event that all free water is not absorbed by the injected glycol, hydrates will form and accumulate in the hydrate detector, causing a pressure drop across it. This indicates the need to reduce the flow of gas or to increase the amount of glycol injected until all free water is removed and the hydrates in the detector are melted.

Condensate stabilization

One of the problems in using low-temperature separation units of both mechanical and glycol injection types is high stock tank vapor loss. This loss is the result of vaporization of appreciable quantities of liquid propane and butane with dissolved methane and ethane, which are liberated when the pressure on the liquid is reduced from the low-temperature separator to storage pressure. When these light ends vaporize in a stock tank, they carry some of the heavier hydrocarbons with them to be burned or lost in the atmosphere. *Stabilization* is a means of removing these lighter hydrocarbons from the liquid present in the bottom of the low-temperature separator with a minimum loss of heavier hydrocarbons. Stabilization results in a larger volume of stock tank liquids available for sale.

A schematic diagram of a glycol injection system with a stabilizer is shown in figure 3.16. The stabilization system consists of a vertical vessel that may be packed with ceramic rings or fitted with trays spaced from 12 to 24 inches apart inside the vessel. The liquid in the lower section of the tower is heated by an indirect heater or steam coils. The cold condensate from the bottom of the low-temperature separator flows directly to the top of the stabilizer or may flow to a flash tank prior to entering the stabilizer. From the flash tank, the gas is routed to become fuel or to be disposed of by other means, and the liquid is taken to the top of the stabilizer. Some of the light hydrocarbons are vaporized and pass from the top of this vessel to the gas sales, fuel, or vent line. The liquid hydrocarbons flow through the packing or down the trays, absorbing some of the heavier gaseous hydrocarbons, which have been vaporized at the bottom of the vessel. At the bottom of the vessel, the heat added from the heater or reboiler vaporizes most of the lighter hydrocarbons. After being cooled, the stabilized liquid flows to storage and the lighter ends flow upward to be reabsorbed or to leave the top of the vessel (fig. 3.17). The amount of additional condensate that can be recovered by stabilization is dependent on the pressure and temperature at which the low-temperature separator is operated and the composition of the gas being processed.

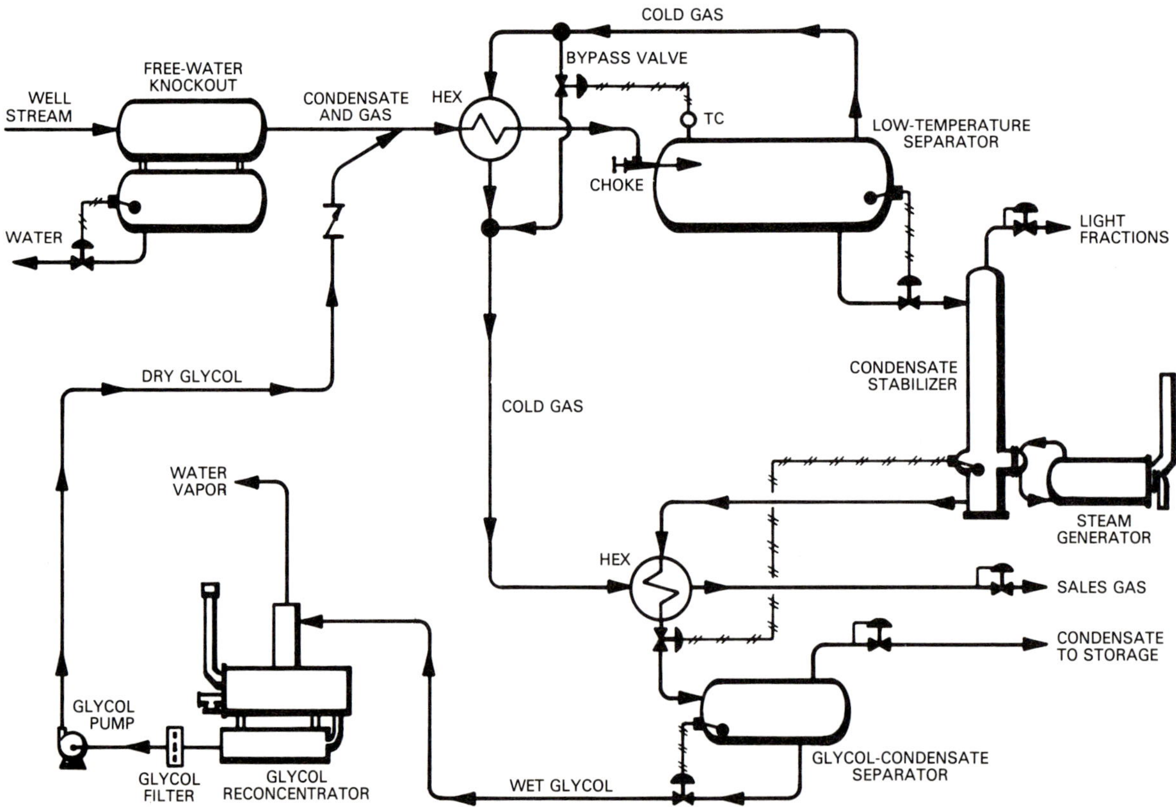

Figure 3.16. Schematic diagram of a low-temperature separation system with glycol injection and condensate stabilization

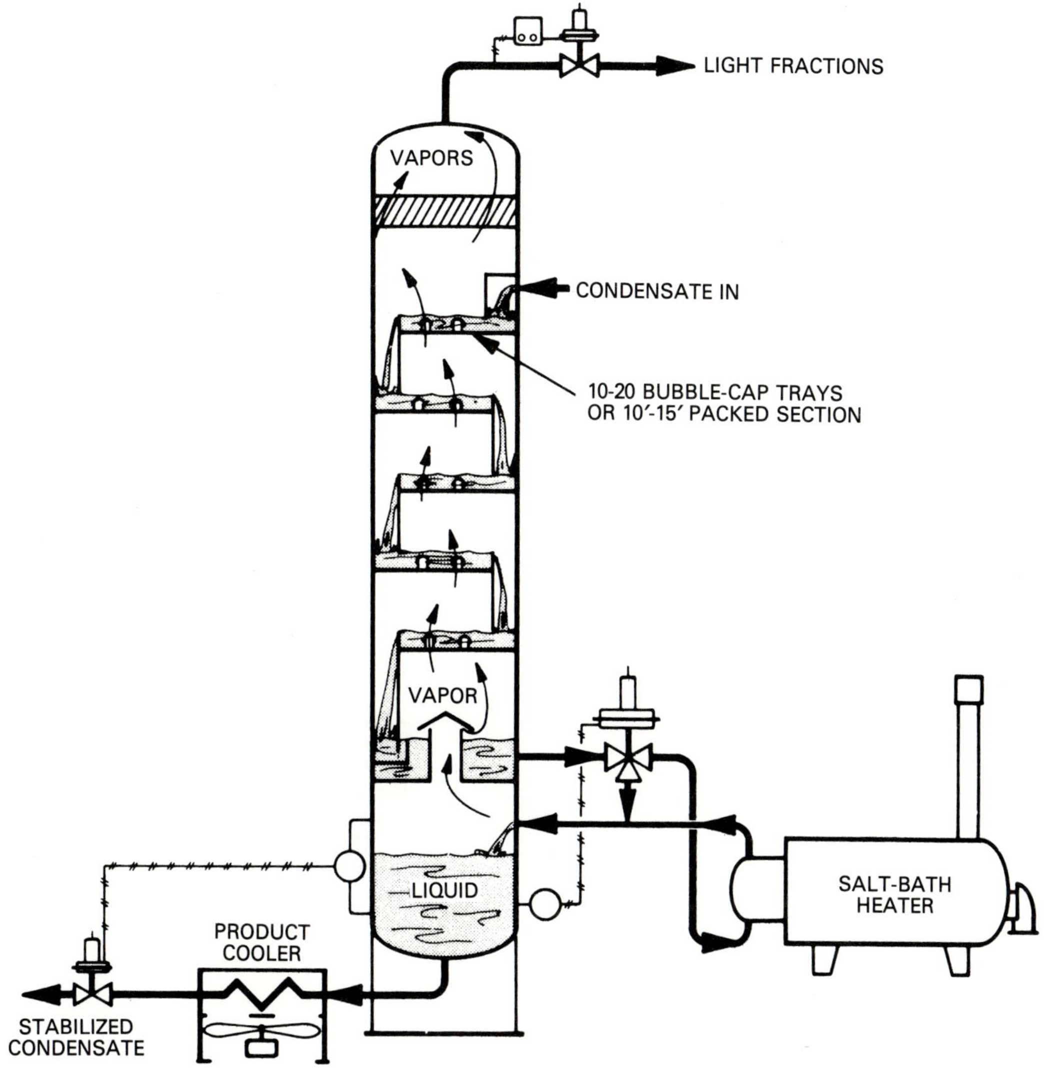

Figure 3.17. Stabilizer with salt-bath heater

Additional hydrocarbon liquid recovery that can be gained by use of low-temperature separation as compared to conventional separation depends on the exact composition of the well stream and on the operating temperature of the low-temperature separator. In the absence of an analysis, the increased recovery may be estimated by assuming that stock tank liquid recovery will be increased about 0.5 bbl per MMscf for each 10°F decrease in separator temperature. For example, if separation at 80°F produces 10 bbl per MMscf, then separation at 20°F will produce an additional 3 bbl per MMscf. An additional increase of about 10 percent might be expected if the low-temperature liquid were then stabilized. For richer gas streams, say in the order of 50 bbl per MMscf, the increased liquid recovery would be about 0.75 bbl per MMscf for each 10°F decrease in separator temperature. These figures are merely an approximation of the actual increased recovery. Figure 3.18 illustrates the effect of separation temperature on liquid recovery.

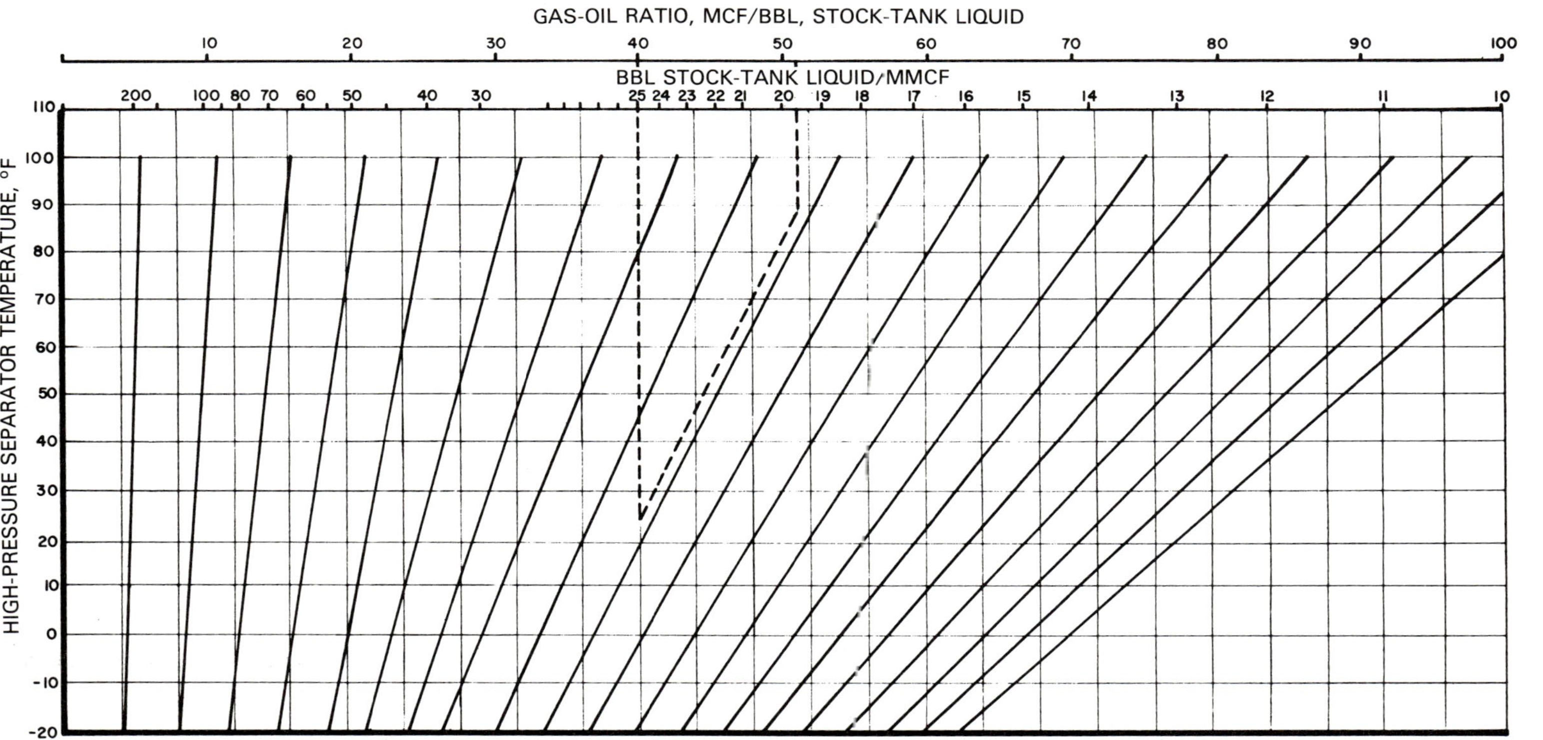

Purpose: To estimate liquid hydrocarbon recovery by low-temperature separation, knowing the recovery or gas-oil ratio by conventional separation.

Method: (1) Locate point of intersection of the known GOR (or Bbl./MMSCF) and the temperature at which the conventional separation test was made. (2) Follow parallel to slanting lines to the intersection of the low-temperature separator temperature. (3) Read vertically from this point to the predicted BBl./MMSCF recovery.

Example: A conventional separation test at 1000 psig showed a GOR of 51 MSCF/stock tank bbl., at a flow rate of 10 MMSCFD at 90 F. If a low-temperature separation unit were used, with wellhead flowing pressure of 3000 psig and separation pressure of 1000 psig, how much additional recovery would result? (Assume 90 F upstream choke temperature, or high-pressure separator temperature.)

Solution: With the separator operating at approximately 25 F., with a 2000 psi differential pressure across the choke, follow the dotted line example on chart. Locate 51 MSCF/Bbl. on GOR scale. Follow it down to the 90 F line. At this intersection draw a parallel line to the nearest slant line until it touches the 25 F line. Then read vertically to 25 Bbl./MMSCF. Since 51 MSCF/Bbl. = 19.6 Bbl./MMSCF, the increased recovery by low-temperature separation would be approximately 25.0 - 19.6 = 5.4 Bbl./MMSCF. At a rate of 10 MMSCFD the increase would be 54 bbl./day. (Note: Well-stream analysis and complete operating data are necessary to predict recoveries accurately. This graph gives a reasonable estimation of recovery.)

Figure 3.18. Effect of separator temperature on liquid recovery

4 Hydrates

Most natural gas contains substantial amounts of water vapor at the time it is produced from a well or separated from an associated crude oil stream. Water vapor must be removed from the gas stream because it will condense into liquid and may cause hydrate formation as the gas is cooled from the high reservoir temperature to the cooler surface temperature. Liquid water almost always accelerates corrosion, and the solid hydrates may pack solidly in gas gathering systems, resulting in partial or complete blocking of flow lines.

Hydrates are solid compounds that form as crystals and resemble snow in appearance. They are created by a reaction of natural gas with water, and, when formed, they are about 10 percent hydrocarbon and 90 percent water. Hydrates have a specific gravity of about 0.98 and will usually float in water and sink in hydrocarbon liquids. Water is always necessary for hydrate formation, as is some turbulence in the flowing gas stream.

Formation of hydrates

The temperature at which hydrates form depends on the actual composition of the gas and the pressure of the gas stream. Therefore, the chart shown in figure 4.1 cannot be completely accurate for all gases, but it is typical for many gases. The chart shows the water content in pounds of water per MMscf of saturated gas at any pressure or temperature. The dotted line crossing the family of curves shows the temperature at which hydrates will probably form at any given pressure. Note that hydrates form more easily at higher pressures. At 1,500 psia, for example, hydrates may form at 70°F, whereas at 200 psia, hydrates will not form unless the gas is cooled to about 39°F. Each curve on the chart shows the water content of a saturated gas at that pressure when the temperature is at any of the various points shown along the bottom of the chart. For example, at 100 psia and 60°F, each 1 million cubic feet of gas would contain about 130 pounds of water vapor. The same gas at 100 psia and 20°F (instead of 60°F) could hold only 30 lb per MMscf. At 0°F, this gas could hold only 13 lb per MMscf.

It can thus be seen that as a gas is cooled, it can hold less water in the vapor form. Therefore, cooling a gas will cause some of the water vapor to condense, with the balance remaining in the gas as water vapor. Pipeline specifications usually require that the water vapor content of natural gas be 7 lb per MMscf or less in order to minimize the problem of hydrate formation in the transmission lines from the field to the ultimate user. In some fields, hydrates form in the tubing and the wellhead valves, necessitating the application of heat down the hole to keep the well from freezing up. In most fields, fortunately, the temperature of the gas at the wellhead is 100°F or more; therefore, the hydrate problem does not usually begin until the gas passes through the Christmas tree.

Ground temperatures

A very important factor in the movement of gas saturated with water vapor is the retention of the heat that is in the gas when it is produced. The temperature is lowered at the wellhead when the gas is expanded through a choke to reduce the pressure and control the rate of flow. After passing the choke, the gas enters the gathering lines, which are cooled by

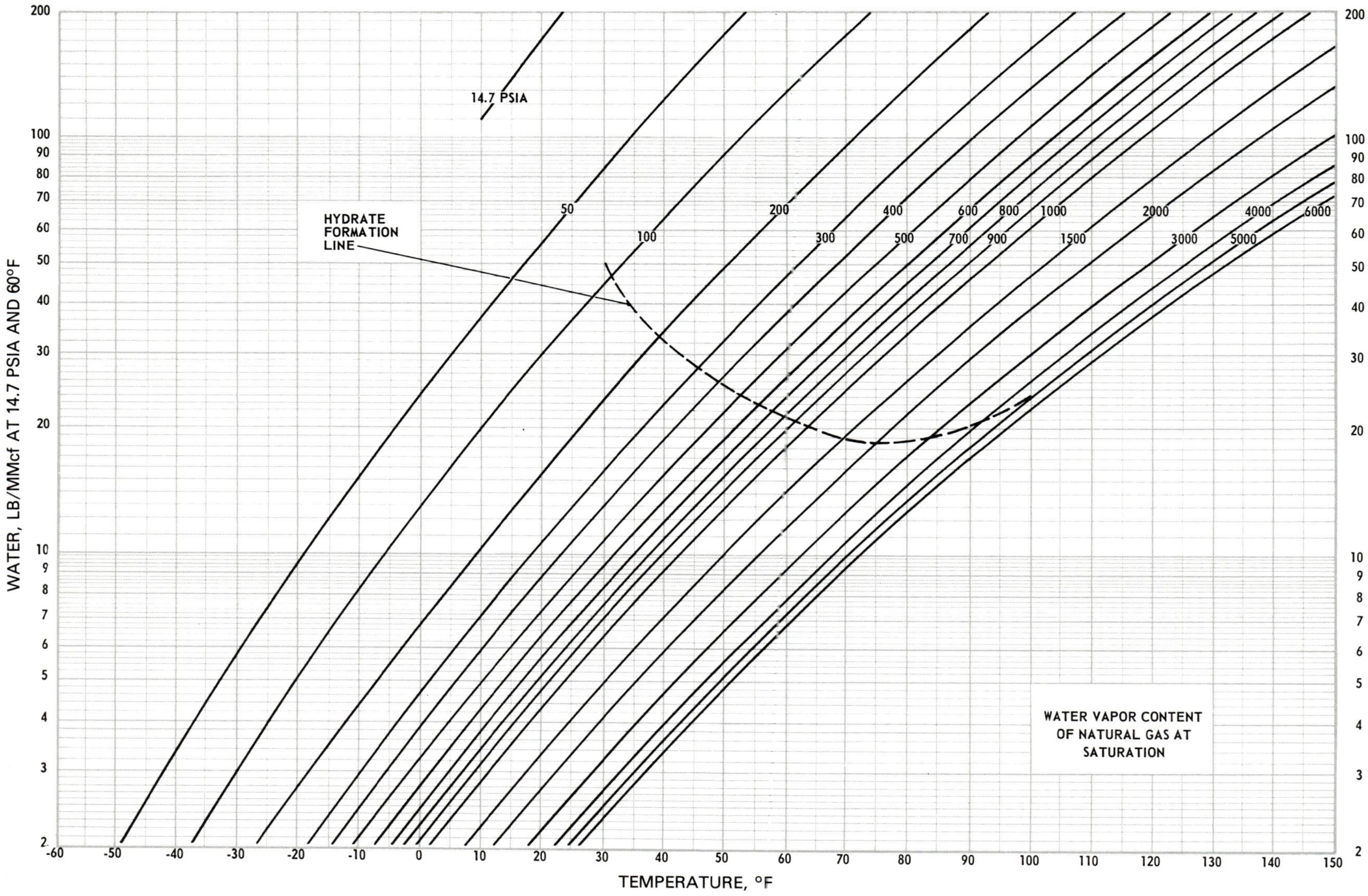

Figure 4.1. Water vapor content of natural gas at saturation

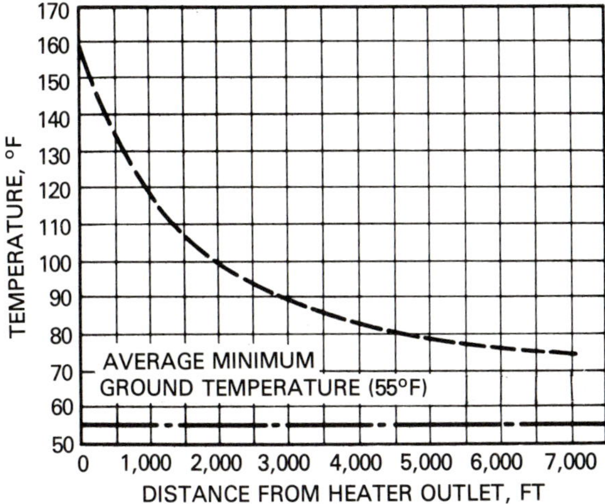

Figure 4.2. Chart of temperature drop in natural gas flow lines

the ground. The effects of the ground temperature are shown in figure 4.2. This curve was prepared from data taken in the Carthage gas field of East Texas. Gas leaving a heater at 160°F had dropped to 75°F by the time that the gas had traveled 1½ miles because of a ground temperature of 55°F. At a higher rate of flow, the gas would travel farther before the formation of hydrates would become a problem. With a line pressure of 900 psi, it may be predicted that hydrates would form at a temperature of about 60°F. Some of the flow lines in this field are as long as 15 miles between well and gasoline plant; in such places the gas has to be reheated before it reaches the plant.

Minimum ground temperatures at a depth of 18 inches in the various gas-producing regions of the Southwest are as follows:

1. Upper Gulf Coast of Texas and Louisiana, 50°F to 56°F
2. Permian Basin, East Texas, and North Louisiana fields, 45°F
3. Hugoton-Panhandle gas areas, 25°F to 30°F

The figures are approximate and represent minimum values. The ground temperatures may be expected to reach higher levels than this during most of the year, especially in the southern part of Texas and Louisiana. On the other hand, it is probable that in the gas-producing areas of the Rocky Mountain states and in western Canada the frost line is below the 18-inch level. Gas lines in those areas are buried at correspondingly greater depths.

Hydrate inhibitors

Ammonia, brines, glycol, and methanol have all been used to lower the freezing point of water vapor and thus prevent hydrate formation in flow lines. Low-volume injectors or pumps are used to feed the inhibitor fluid into the gas flow system. Methanol and glycol are the inhibitors most widely used; however, they are expensive. Usually methanol and glycol are used when hydrate problems are so rare that the installation of a heater or dehydration equipment is not economically feasible. For example, gas-producing operations may be carried on without hydrate trouble except for two or three days in each year. In such cases, injection of inhibitors may be preferred to investment in additional equipment.

Flow-line heaters

In a great majority of situations in which dehydration of gas is not economical, yet some measure must be adopted to control hydrate formation, the method used is the application of heat. The reasons are obvious. Initial investment is not excessive; fuel is conveniently available; and heaters operate with a minimum of attention. When it is desirable to handle the gas by means of temporarily holding the temperature above ground temperature, heaters seem to offer the best solution. For long-distance transmission, the gas will eventually reach ground temperature, and water will eventually have to be removed.

Indirect heaters with water bath

Most natural gas is produced at relatively high flow line pressures, ranging from 1,500 to 10,000 psig, and then transferred to pipelines operating at 1,200 psig and lower. Heat must almost always be used to compensate for the natural refrigeration of the gas, which is caused by this pressure reduction. An indirect heater is the most widely used type because it is simple, economical, and, if properly sized, a trouble-free piece of equipment. The heater consists of three basic parts: (1) the heater shell, which is a thin-walled horizontal vessel having removable flanged covers at both ends; (2) a removable fire tube and burner assembly mounted on the lower portion of one of the end covers; and (3) a removable coil assembly mounted on the upper portion of the opposite end cover. The shell and fire tube are designed to withstand only atmospheric pressure, whereas the coil assembly is usually designed to withstand shut-in wellhead pressure.

Arrangement of the equipment of an indirect heater is illustrated in figure 4.3. The high-pressure fluid is introduced to the heater through a choke located on the inlet of the heater, and an expansion takes place immediately inside the heater in the water bath. As the fluid passes through the steel coils, it is heated to a suitable outlet temperature. In operation, the heater shell is filled with a liquid, usually fresh water, completely covering the fire tube and the coil assembly. The water bath is heated by the fire tube, and the coil assembly is heated by the water. Because the water serves as a medium of heat transfer, the heater is called indirect. The capacity of such a heater is related to the Btu rating of the burner assembly and fire tube and the effective area of the well stream coil.

As illustrated, the coil of an indirect heater is located directly above the fire tube. Both the coil and the fire tube are immersed in the water bath. Other liquids may be used, but ordinarily the liquid is water. Fundamentally, the fire tube is the heat source, and the coils are the points of heat absorption. Hot water has a lower density than cold water and, in a closed vessel, will rise, displacing the cold water as the latter settles to the bottom of the

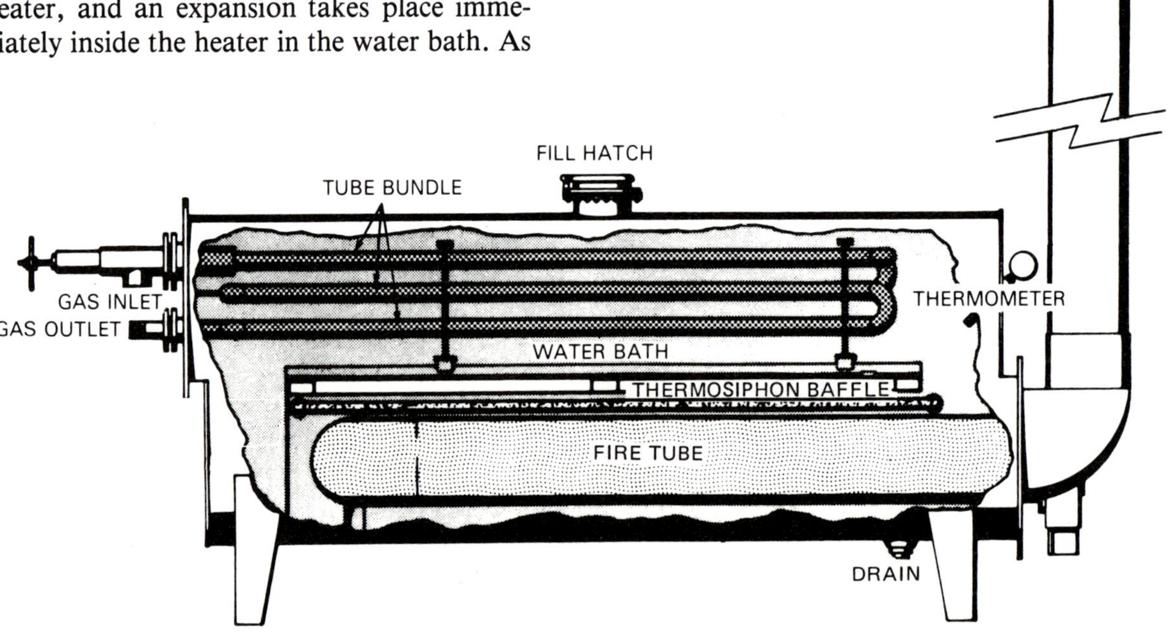

Figure 4.3. Cutaway view of an indirect heater

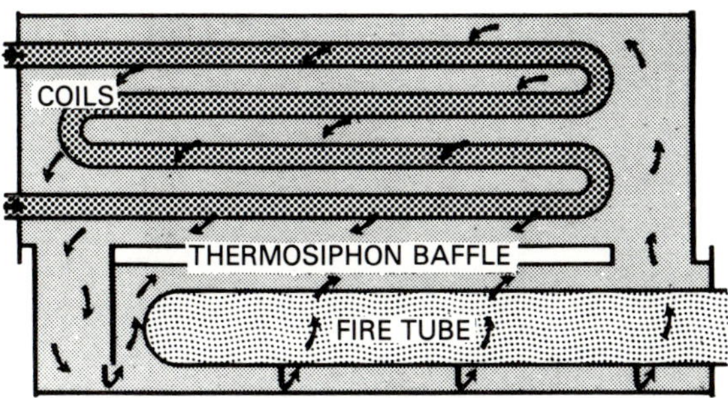

Figure 4.4. Thermosiphon baffle arrangement for an indirect heater

vessel because of gravity. For that reason the coils are located above the fire tube. Heat travels from the fire tube to the water immediately surrounding it and causes thermal currents in the water. The direction of these currents can be controlled by the installation of a thermosiphon baffle (fig. 4.4).

The advantages of directionally controlled thermal currents inside an indirect heater are twofold.

1. The positive velocity of the water bath past the fire tube surface carries away the fire tube heat at a faster rate and reduces the probability of steam generation, thereby reducing and generally eliminating scaling of the fire tube.
2. The overall heat transfer efficiency of the coil, which is located in the upper part of the heater, is increased because of the increased velocity over the surface of the coils.

Coil design is important, especially so because the coils of an indirect heater are often manifolded together into bundles, with outside piping tying the two bundles together according to requirements.

Most indirect heaters use natural gas as fuel (fig. 4.5). The inlet gas, usually taken from the first-stage separator, is heated in a small coil immersed in the water bath. Then the gas is regulated to 25 to 40 psig and passed through a safety fuel-gas scrubber, a strainer, the thermostatically controlled valve, the final low-pressure regulator, and the burner assembly. Fuel for the pilot light is taken upstream of the thermostatically controlled valve. An important safety factor is that the main gas valve is prevented from opening if the pilot flame fails for any reason.

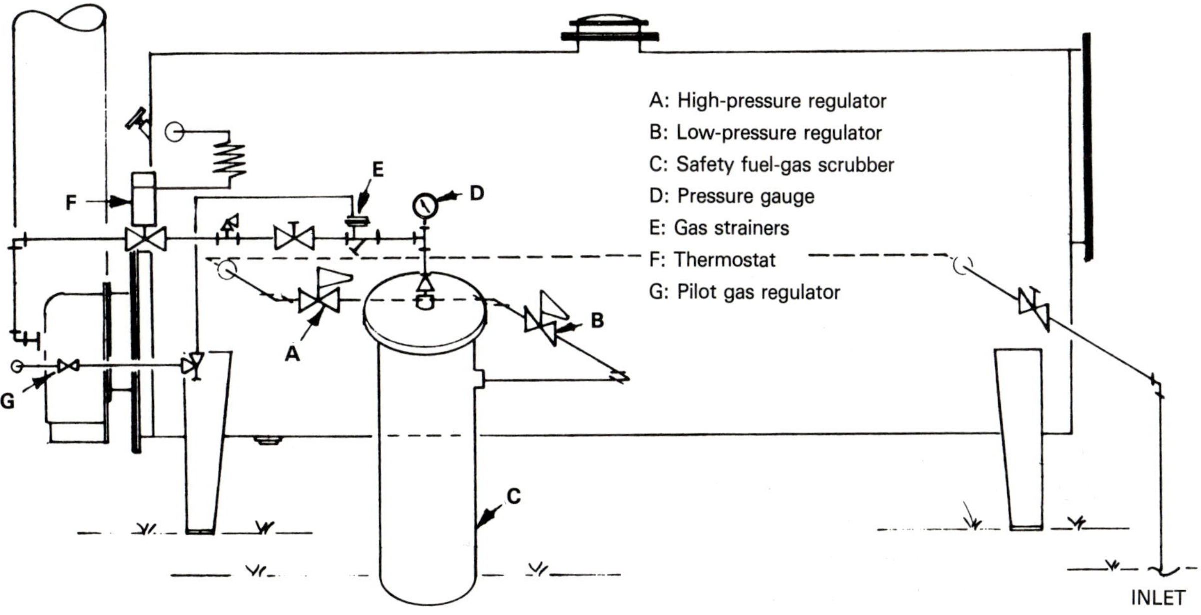

Figure 4.5. Schematic hookup of fuel-gas system for indirect heater

When indirect heaters must be located in potentially hazardous areas, the burner assembly and sometimes the stack assembly are equipped with flame arrestors. These devices permit the heater to draw in sufficient air for proper combustion but to prevent explosions or flames within the fire tube from igniting the surrounding atmosphere.

The following features of a flame arrestor are essential:

1. A bank or honeycomb of small linear passages acts as a mass heat exchanger. These passages are staggered to provide protection and safety.
2. The case holding the arresting element must also withstand a fire tube explosion and prevent the element from being blown out of the case.
3. The element of the arrestor must be sized to provide sufficient open area for air entrance so that the heater can develop its full Btu rating without requiring forced-draft equipment.

Research has indicated that most heater coil failures result from a combination of corrosion and erosion. Corrosion attacks the surface of the metal, and erosion wipes away the products of corrosion, leaving the metal face clean and therefore more receptive to new corrosion. It is a vicious and continuous cycle. It has been determined that the rate of corrosion and erosion in the return bends is greater in every case than in the straight-pipe coils attached to the return bends.

Through field experience and laboratory analysis, manufacturers searched for a way to build into the indirect heater coils a warning signal. It was found that some refineries made a practice of using safety drilling tees, ells, return bends, pipe, and valves. Safety drilling was a simple process of drilling a small hole on the outside of the return bend approximately one-half through the thickness of the metal at a point where erosion and corrosion were likely to be concentrated (fig. 4.6). Such safety drilling provided a place that would issue a warning leak of gas or liquid when approximately half of the metal thickness had corroded or eroded away.

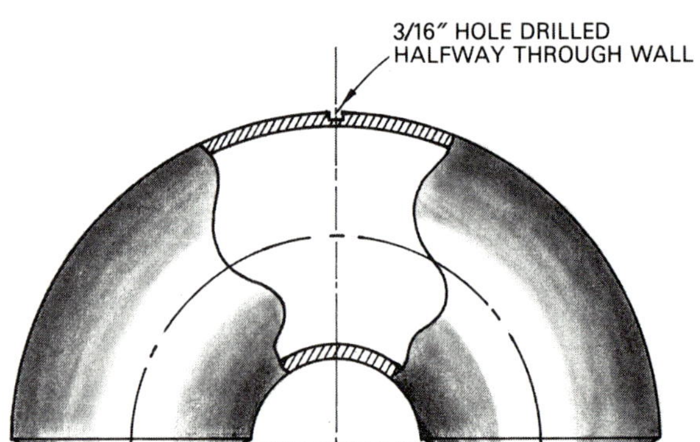

Figure 4.6. Safety-drilled return bend for an indirect heater

In a long-nose choke, the choke body is extended so that the choke orifice may be located within the indirect heater bath (fig. 4.7). Thus a water-saturated gas can be expanded to a temperature below the hydrate-forming temperature of the gas, yet no freezing or plugging of the coil will occur. The use of a long-nose choke results in a greater mean temperature difference between the coil fluid and the water bath. Since the coil area required to perform a particular heat exchange job is inversely proportional to the mean temperature difference, the extra cost for the long-nose choke over the standard choke can, in most cases, be paid for in lower coil area requirements.

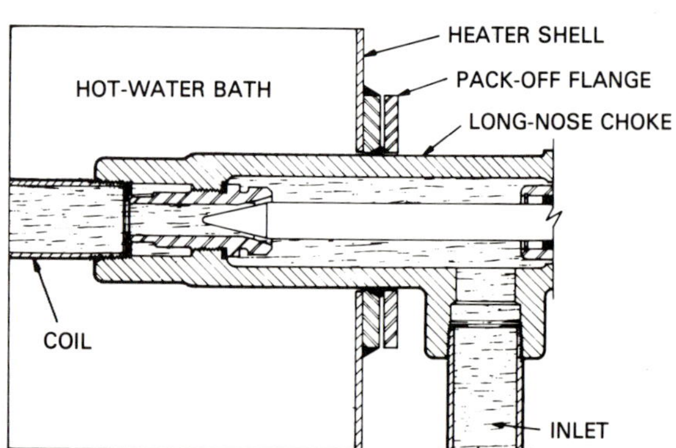

Figure 4.7. Long-nose choke arrangement for an indirect heater

Indirect heaters with other bath solutions

While most indirect heaters use a water bath, others with different liquids and solutions warrant mention.

1. Calcium chloride water solution with inhibitor is used in some heaters to prevent freezing of the water bath. Fouling of the coils will occur, however, if deposition of calcium chloride occurs.
2. Some heaters use a glycol-formulated antifreeze with inhibitor in the bath to reduce the heat transfer rate from the bath to the coil. If the bath is half glycol and the bath temperature remains unchanged, the heat transfer coefficient may be reduced as much as 20 percent. However, the bath temperature may be increased to 200°F at sea level pressure without boiling, compensating somewhat for the loss in heat transfer coefficient with a higher mean temperature difference.
3. Heaters may have a steam bath, in which the temperature is 245°F at 15 psig and sea level atmospheric pressure. Smaller coil area is possible because of (1) the greater mean temperature difference and (2) a larger heat transfer coefficient. Insulation for such a heater must be considered, and those stamped with the ASME code should be employed. In some areas, the ASME vessel is mandatory.
4. Heaters with a molten salt bath are generally employed in process plants, dehydration units, and hydrocarbon recovery plants. They are not applicable to ordinary gas- or oil-producing operations because the high melting temperature—about 600°F—of the salt bath is considerably above maximum temperature requirements for the usual producing operations.

5 Dehydration of natural gas

The process whereby water vapor—and certain other vapors as well—are removed from natural gas by either absorption or adsorption is termed dehydration. The removal of water vapor by bubbling the gas countercurrently through certain liquids that have a special attraction or affinity for water is known as *absorption.* Removal by making the gas flow through a bed of granular solids that have an affinity for water is called *adsorption;* the water is retained on the surface of the particles. The vessel in which either absorption or adsorption takes place is called a *contactor,* or *sorber.* The liquid or the solid having affinity for water and used in the contactor in connection with either of the processes is called the *desiccant.*

The *dew point* of natural gas is that temperature at any specified pressure at which it is saturated with water vapor. Being saturated means that the gas contains in vapor form all the water possible at the specified pressure and temperature.

Two major types of dehydration equipment are in use at this time, namely, the *liquid desiccant* dehydrator and the *solid desiccant* dehydrator. Each has its special advantages and disadvantages and its own field of particular usefulness. Practically all of the gas moved through transmission lines is dehydrated by one or the other of these two methods.

Dew point depression

Hydrates do not form in a gas line unless the gas is saturated and contains still more water that, since it cannot be absorbed, takes the form of free water. At any specified pressure, hot gas takes more water vapor to reach the saturation point than does cool gas. This means that cool gas is saturated and also has some free water; it will absorb all of the free water when heated sufficiently at the same pressure. If heated above this point, cool gas will not only take up all the free water as water vapor and so prevent hydrate formation, but also will be undersaturated—that is, it will be capable of absorbing more water vapor than the gas is holding. For example, gas at 500 psia and 60°F at the saturation point contains 30 lb of water per 1 MMcf. The dew point of this gas is 60°F. Suppose this gas is going to be moved to New York in a transmission line with a temperature of 20°F. The saturation point will then be 7 lb of water per 1 MMcf. The original 30 lb of water, if left in the gas, will then exist in the form of 7 lb of water vapor and 23 lb of free water per 1 MMcf, if the pressure remains the same. This free water is a potential source of hydrates that may freeze and plug the line. Suppose that the gas is processed in a dehydration unit and the dew point is depressed 50°F. No free water will exist in the gas until the temperature is lowered more than 50°F from the original temperature of 60°F or until the temperature goes to 10°F or lower. Gas at 500 psia and 10°F contains about 5 lb of water vapor per 1 MMcf. Since this gas originally contained 30 lb of water vapor per MMcf, it will be necessary to remove 25 lb of water from each 1 MMcf in order to depress the dew point 50°F. In principle, this depression is the job of the dehydration unit. The problem presented has been purposely stated in oversimplified form to establish the principle of operation. Actual operating problems are not so simple.

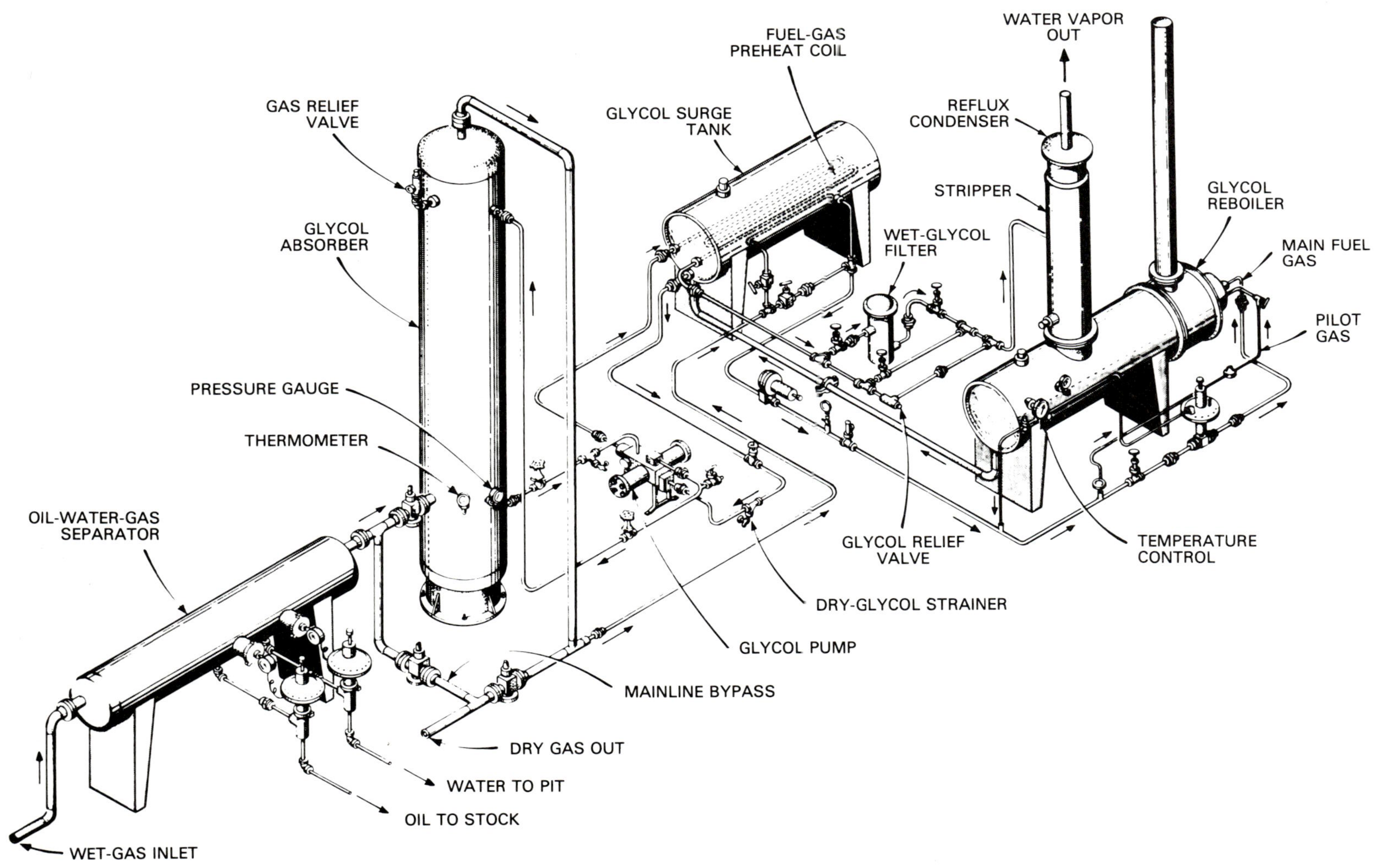

Figure 5.1. Flow diagram of a liquid desiccant unit

Liquid desiccant dehydrators

The desiccant used in a liquid dehydrator is usually a solution of one of the glycols—diethylene glycol (DEG) or triethylene glycol (TEG). Triethylene glycol is superior as a desiccant. It is more easily regenerated to 98-99 percent concentration, has a higher decomposition temperature (about +40°F), and is subject to lower vaporization losses.

The following definitions apply to the process description that follows and to the flow diagram in figure 5.1.

Wet gas is gas containing water vapor prior to contacting glycol in the absorber.

Dry gas is gas leaving the absorber after contacting glycol.

Desiccant is a drying or dehydrating medium; in the process description, the desiccant is a triethylene glycol-water solution

Lean solution is a glycol-water solution whose glycol concentration ranges from 95 to 99 percent by weight. A lean solution can be the solution passing from the reboiler via the pump to the sorber, a reconcentrated solution, or TEG supplied in sealed drums.

Rich solution is a water-rich solution whose glycol content is less than 95 percent by weight or glycol solution that has contacted wet gas in the sorber.

Natural gas at line temperature and pressure enters near the bottom of the absorber (figs. 5.1 and 5.2) and rises through the column, where it is intimately contacted by a lean glycol solution flowing downward across bubble trays. Here the gas gives up its water vapor to the glycol. Leaving the top tray, the gas passes through mist-extractor elements, sweeps the glycol-cooling coils located in the upper end of the absorber, and passes to the pipeline. A small quantity of this dry gas is withdrawn from the absorber discharge for use as fuel and instrument gas.

The lean glycol solution enters at the top of the absorber and flows through coils where it is cooled by the dehydrated gas. From the cooling coils, the glycol is discharged in intimate contact with the ascending gas; this action dehydrates the gas and dilutes the glycol at the same time. The dilute solution collects in the base of the absorber, from which point it is discharged to the reconcentrator or reboiler. Enroute, the solution is heated in a coil immersed in the glycol surge tank and then discharged into the stripper, which is normally a packed column. The reconcentrated glycol solution accumulates in the reboiler, where it reaches maximum temperature; overflows into the glycol surge tanks, where it is partially cooled by heat exchange to the dilute glycol in the coil; and flows by gravity to the pump suction. The pump discharges the concentrated solution into the cooling coil in the absorber, thus completing the cycle. Figure 5.3 pictures a small glycol unit.

A dilute glycol solution can be reconcentrated simply by heating it in an open pot to drive off the water as steam. However, a substantial quantity of glycol is thereby driven off with the water, and the loss is sufficiently large to make the use of a simple boiler economically undesirable. Since water and triethylene glycol have widely varying boiling points (212°F and 549°F), the two substances can be separated easily by fractional distillation. This is accomplished in the packed column, or stripper, mounted atop the reboiler. Within this column, water-rich vapor from the reboiler rises in intimate contact with descending glycol-rich liquid from the absorber. Between the two phases there is a continuous exchange of heat and materials, causing the glycol vapor to condense and liquid water to vaporize as equilibrium is approached. At the top of the column the vapor is virtually pure water, while there is very little water in the glycol at the bottom. By condensing water in the top of the column with an atmospheric heat exchanger or a reflux cooling coil through which the cool,

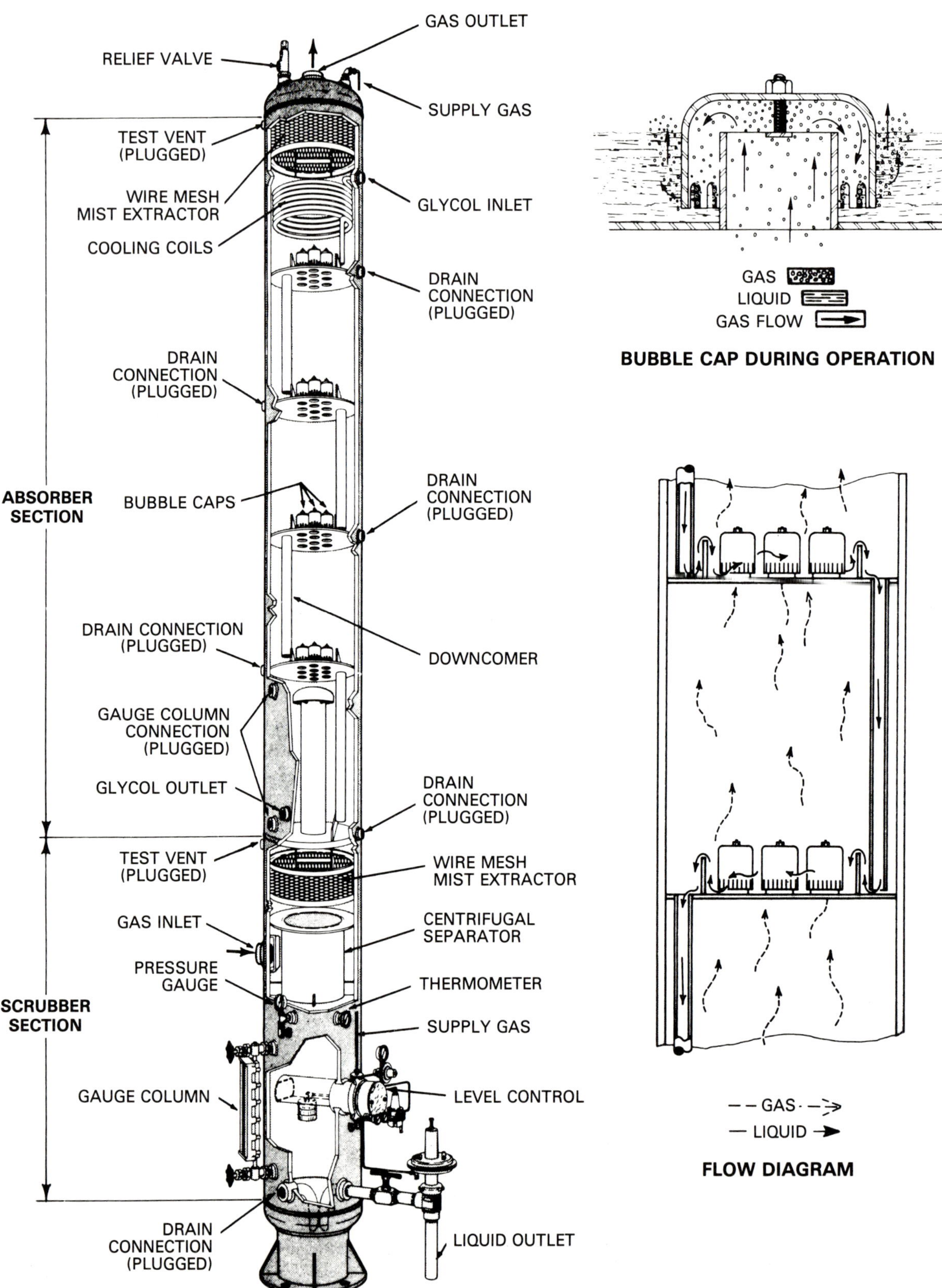

Figure 5.2. Glycol absorber tower with scrubber section

Figure 5.3. Glycol liquid desiccant dehydrator unit

rich glycol circulates before entering the still proper, sufficient liquid, known as *reflux,* is provided for proper fractionation. As long as the temperature at the top of the column ranges from 210°F to 212°F, glycol losses are minimized.

In order to understand the varying conditions under which most liquid desiccant dehydrators operate, it is necessary to consider the effect of four major operating variables: (1) gas pressure, (2) gas temperature, (3) solution rate, and (4) solution concentration. At constant temperature, the lower the pressure, the higher the water content of the inlet gas will be. At constant pressure, the higher the temperature, the higher the water content of the inlet gas will be. Over a normal pressure range up to 1,200 psig, about 2 gallons of glycol must be circulated for every pound of water removed at the 55°F dew point depression. This quantity is based on equilibrium conditions and a 95 percent glycol solution. Greater dew point depressions can be attained by increasing the circulation rate to the upper limit of the design capacity of the equipment. Also, as the circulation rate exceeds 7 gallons per pound of water content, the effect of dew point depression is greatly diminished. In general, as the glycol concentration becomes higher, the dew point depression becomes greater.

Whereas glycol concentrations of 95 to 96 percent will give dew point depressions of 55°F, concentrations of 99 percent will give depressions of 65°F, assuming that the circulation rate remains constant. A reboiler temperature of about 350°F will give a glycol concentration of 95 to 96 percent, and a reboiler temperature of about 375°F will give a glycol concentration of 97 to 99 percent. Both of these values apply at atmospheric pressure. Some manufacturers are marketing high-concentration glycol regeneration units, which permit glycol dehydrators to compete with dry desiccant units for low dew points. Dew point depressions of 130°F to 140°F are now possible with glycol dehydrators using triethylene glycol and the high-concentration glycol regenerators. TEG concentrations of 99.1 to 99.995 percent can be obtained with the stripping gas feature.

Most operating problems encountered in the operation of glycol dehydration plants are related to contamination of the glycol by foreign materials and to the breakdown of the glycol at elevated temperatures. The first place in the system for removal of contaminants is the inlet wet gas scrubber. This vessel should be located as close as possible to the absorber. Its function is to remove water, hydrocarbon condensate, crude oil, lubricating oil, and any pipeline dust or dirt that comes along with the gas. The vessel should be of adequate size and be maintained in proper working order; if light oils and condensate are carried into the absorber, they eventually find their way to the reconcentrator, where they are likely to flash. Flashing of these elements may cause fracturing of the packing in the still and blow glycol and hydrocarbons out of the top, resulting in a serious fire hazard.

Heavy oils will eventually find their way to the surge tank, where they must be drained off. Some oils form an emulsion with the glycol. When this emulsion reaches the regenerator, light fractions of the oil are driven off overhead with the water. In several cases, fires have been started when these vapors have come into contact with the fire in the reboiler. Such fires can be prevented by carrying the vapors away from the top of the still column through a downward sloping line into a remotely located separator that can draw off the water separately from the gasoline. Care must be taken to prevent this line from freezing.

A glycol filter is usually installed to remove dirt, scale, rust, and reaction materials. This filter is usually of the waste-pack or cartridge type, and can be removed and replaced while the plant is in operation. A type of filter sometimes used has an element of cloth, fiberglass, or cotton in the form of a replaceable sock, surrounded by a stainless steel wire mesh. It is located upstream of the motor valve controlling the liquid level in the absorber, since gas released because of the pressure drop through the motor valve would nullify its operation. The filter in this location must be kept clean. A dirty filter raises the back-pressure and the liquid level in the absorber, causing the motor valve to open wider until the pressure drop is taken across the filter instead of the motor valve, with a resultant bursting of the filter sock.

Glycols are difficult liquids to contain and to pump. The liquid continually leaks around packing and soon spreads around the entire pump area if not properly handled. The tendency of inexperienced operators is to tighten the packing too tightly. The leak problem can be overcome by using a good packing and by providing a drainage trough around the pump, leading to a central sump from which the glycol can be gathered and returned to the main reservoir. Metallic packing with a soft babbitt center combined with graphite and mica has proved successful. This self-lubricating packing limits the leakage to one or two drops every several minutes.

The reboiler is usually direct-fired or steam-heated, with the glycol level maintained by an overflow pipe to the surge tank or by a liquid-level controller. When level controllers are used, the float is normally located in the reboiler, and the control valve is on the absorber. A time lag is thus introduced between the opening of the valve and the raising of the level in the reboiler. Failure of the controller or the valve will result in a low liquid level in the reboiler, resulting in a possible burning out of the fire tube. The overflow system, with the surge tank directly below the reboiler, minimizes this hazard simply by ensuring a minimum liquid level. In cases where fuel for the reboiler is not constant or in areas of high winds, operators usually equip the units with a pump shutdown device in the event of flame failure. This device prevents the glycol from circulating when the reboiler is out of service and prevents the system from building water and losing glycol.

The still column, operating at the highest temperature in the system, shows the effects of the highest rates of corrosion. Reboiler corrosion can be minimized by keeping the system free of contaminants and operating at a minimum temperature. One method of keeping the temperature down is to elevate the reboiler, providing the necessary gravity head to return the glycol to storage, and to operate the reboiler at atmospheric pressure. Another remedy is to reduce the capacity of the reboiler so that the temperature is raised and lowered quickly, minimizing the time in which the glycol is at the elevated temperature and subject to decomposition. Some operators use stainless steel construction, but most designs include a metal thickness ample enough to take care of corrosion.

Another problem encountered in dehydration plant operation is foaming. Glycols foam with hydrocarbon condensate emulsion, some corrosion inhibitors, and corrosion products. Removal of these materials before they enter the absorber is the most reliable preventive measure. In some cases, defoaming agents are effective. Foaming problems require individual attention, since each case is unique and no set formula can be applied.

The loss of glycol is one of the most common and consistent problems facing the operator of a dehydration plant. Some glycol losses are normal and unavoidable; the best of systems will lose about 0.1 gallon per 1 MMcf of gas passing through the absorber. The losses may be slightly higher than this at very low pressures or very high temperatures. Such unavoidable loss is the small amount of glycol that vaporizes as the gas bubbles through the glycol in the absorber or that leaves in the form of mist from the top of the absorber.

Another normal and unavoidable loss is that small amount of glycol that vaporizes with the water in the reconcentrator still column.

It is not economical to reduce this loss beyond a certain point because of the high cost of elaborate refluxing systems. Glycol can be lost through leaks from the bottom or the chimney tray into the integral scrubber section of an absorber tower. This lost glycol will be discharged with the condensate or water and is difficult to detect. A leak in the glycol-gas heat exchanger can cause glycol losses and is equally difficult to locate. Glycol may be lost through the absorber outlet because of a foaming condition, usually caused by condensate in the gas, compressor oils, or high concentrations of solid particles. Many effective anti-foaming agents are available, but a high-quality inlet gas scrubber is usually the best solution to the problem. Foaming may also occur in the reconcentrator and cause glycol losses through the still vent line. Plugging of the still column packing is another cause for glycol losses; this is almost always caused by deposition of solid particles in the packed still column or by breakage and compaction of the still column packing material.

Still column packing, which is normally a ceramiclike material, can be broken if a slug of cold liquid hits the relatively hot packing. Repeated heat shocks will ruin the packing in time. If a slug of free water or light condensate should enter the absorber, it would soon be dumped to the reconcentrator, where it would vaporize or flash and cause excessive vapor velocity through the still column. This would carry out any glycol in the column and, at the same time, could lift and drop the packing and damage it.

Improper start-up procedures can cause glycol losses. The sales gas line downstream of the absorber should be pressured up very slowly or the absorber bypassed before putting the system on stream. A low pressure in this sales line can cause excessively high gas flow rates through the absorber for a short period of time and cause all the glycol in the absorber to be blown down the line. Glycol can be lost if the trays in the absorber are not filled before start-up.

Solid desiccant dehydrators

When the highest possible dew point depression is required, the solid desiccant dehydration system is the most effective type. It is not uncommon to process gas through these systems with a resultant residual water vapor in the outlet gas of less than ½ lb per MMscf. In the average system, this amount might correspond to a dew point of −40°F. Dehydrators of this type are manufactured as packaged units ranging in capacity from 3 to 500 MMscf/d, with design pressures of from 300 to 2,500 psig. Solid desiccant units find their greatest application in gas transmission line systems.

The essential components of a solid desiccant dehydrator installation (fig. 5.4) are—

1. an inlet gas stream separator, usually a filter separator;
2. two or more adsorption towers (adsorbers or contactors) filled with a granular gas-drying material;
3. a high-temperature heater to provide hot regeneration gas for drying the desiccant in the towers;
4. a regeneration gas cooler for condensing water from the hot regeneration gas;
5. a regeneration gas separator to remove water from the regeneration gas stream; and
6. piping, manifolds, switching valves, and controls to direct and control the flow of gases according to process requirements.

The following terms apply to the technology of solid desiccant dehydrations:

Wet gas is gas containing water vapor prior to flowing through the adsorber towers.

Dry gas is gas that has been dehydrated by flowing through the adsorber towers.

Regeneration gas is wet gas that has been heated in the regeneration gas heater to temperatures of 400°F to 460°F. This gas is passed through a saturated adsorber tower

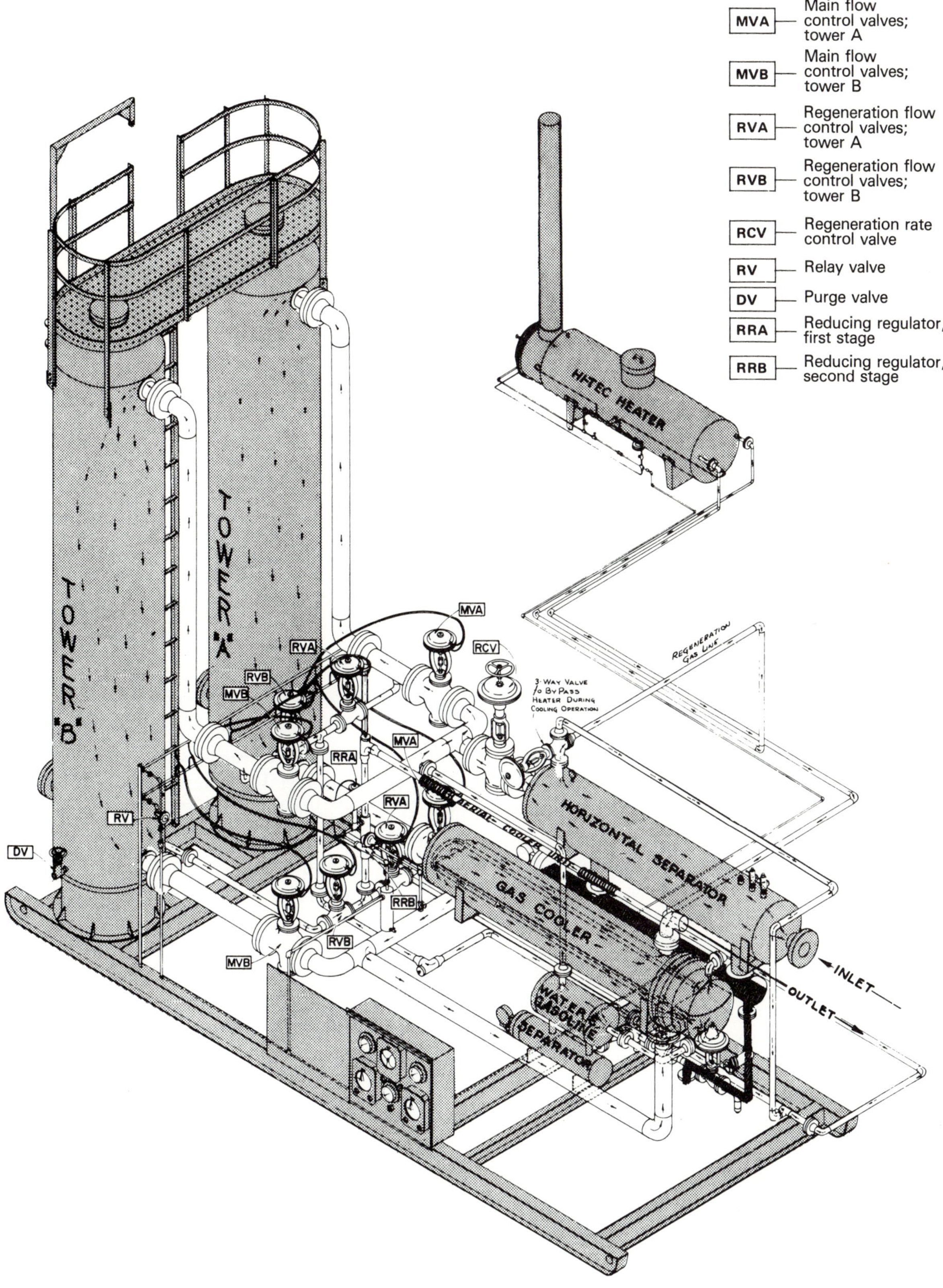

Figure 5.4. Two-tower solid desiccant dehydration unit

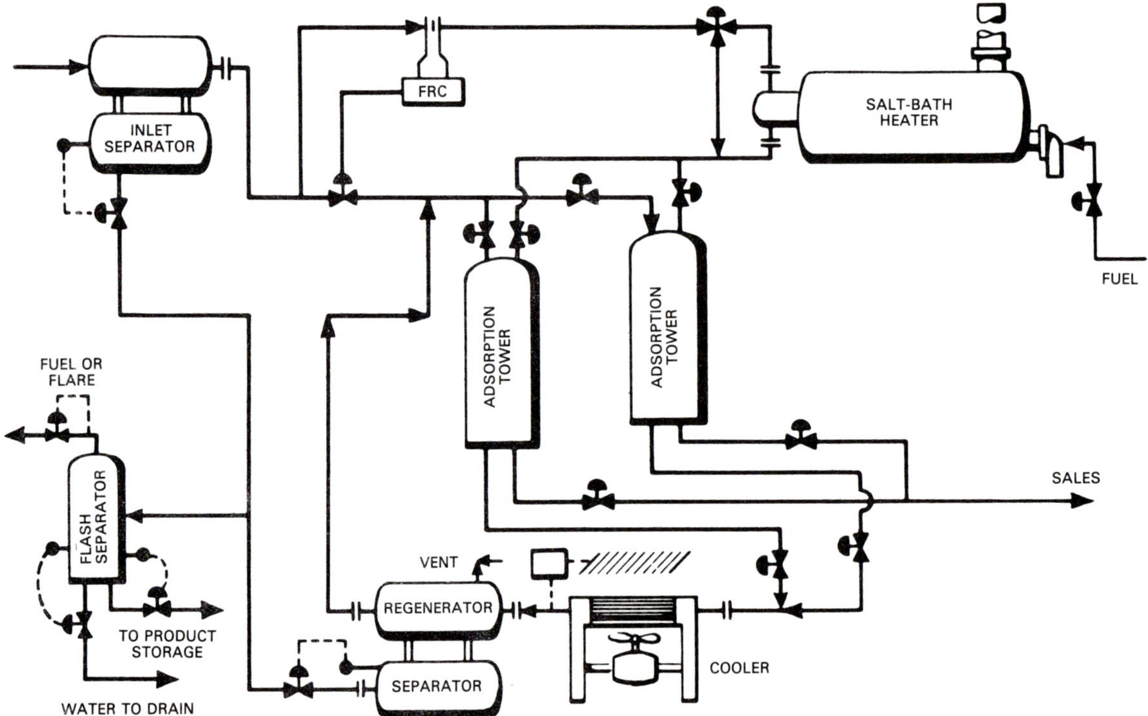

Figure 5.5. Flow diagram of a two-tower solid desiccant dehydration unit

to dry the tower and remove the previously adsorbed water.

Desiccant is a solid granulated drying medium that has an extremely large effective surface area per unit weight because of a multitude of microscopic pores and capillary openings. A typical desiccant may have as much as 4 million square feet of surface area per pound.

The term *adsorption* refers to the effect that natural forces have on the surface of a solid in tending to capture and hold vapors and liquids on its surface. Adsorption processes, as opposed to absorption processes, do not involve chemical reactions. Adsorption is purely a surface phenomenon. All solids adsorb water to some extent, but their efficiency varies primarily with the nature of the material, its internal connected porosity, and its effective surface area. In most dehydration systems, activated alumina (bauxite) or a silica gel desiccant is used. Adsorbents are specific in nature, and not all adsorbents are equally effective. Different molecules are attracted to adsorbents at different rates. Because of this, adsorbents are capable of separating materials preferentially, in either gaseous or liquid phases. The separation is accomplished by passing the stream to be treated through the tower packed with adsorbent. The degree of adsorption is a function of operating temperature and pressure; adsorption, up to a point, increases with pressure increase and decreases with temperature increase. A bed may be regenerated either by decreasing its pressure or by increasing its temperature.

Adsorber towers are made ready for new adsorption cycles by increasing the bed temperature and passing a stream of very hot gas through it. The hot natural gas not only supplies heat but also acts as a carrier to remove the water vapor from the bed. After the bed is heated to a predetermined temperature, it is cooled by the flow of unheated gas and thus made ready for another cycle.

Figure 5.5 is a flow diagram of a two-tower solid desiccant dehydration unit. The wet inlet gas stream first passes through an efficient inlet separator where free liquids, entrained mist, and solid particles are removed. This part of the system is very important, since free liquids may damage or destroy the desiccant bed and solids may plug it. If the plant happens to be downstream of an amine unit or a compressor station, a filter inlet separator should be used.

At any given time, one of the towers will be on stream in the adsorbing cycle, and the other tower will be in the process of being regenerated and cooled. Several automatically operated switching valves and a controller route the inlet gas and regeneration gas to the proper tower at the proper time. Typically, a tower will be on the adsorb cycle for 4–12 hours, with 8 hours being the most common time cycle. The tower being regenerated will be heated for about 6 hours and cooled during the remaining 2 hours. Large-volume systems may have three towers (fig. 5.6). At any given time, one tower will be in the adsorption cycle, one tower will be in the heating cycle, and the remaining tower will be in the cooling cycle.

As the wet inlet gas flows downward through the tower on the adsorption cycle, all of the adsorbable gas components are adsorbed at different rates. The water vapor is immediately adsorbed in the top layers of the bed. Dry hydrocarbon gas components (ethane, propane, butane, etc.) passing on down through the bed are also adsorbed, with the heavier components displacing the lighter components as the cycle proceeds. As the upper layers of desiccant become saturated with water, the lower layers begin to see wet gas and begin adsorbing the water vapor, displacing the previously adsorbed hydrocarbon components. For each component in the inlet gas stream, there will be a section of bed depth, from top to bottom, where the desiccant is saturated with that component and

Figure 5.6. Three-tower solid desiccant dehydration units for a transmission line

where the desiccant is just starting to see that component. The depth of bed from saturation to initial adsorption is known as the *mass-transfer zone.* This is simply that zone or section of the bed where a component is transferring its mass from the gas stream to the surface of the desiccant.

As the flow of gas continues, the mass-transfer zones move downward through the bed, and water displaces all of the previously adsorbed gases until finally the entire bed is saturated with water vapor. When the bed is completely saturated with water vapor, the outlet gas will be just as wet as the inlet gas. Obviously, the towers must be switched from adsorb cycle to regeneration cycle before the bed has become completely saturated with water.

Regeneration gas is supplied by taking a portion of the entering wet gas stream across a pressure-reducing valve that forces a portion of the upstream gas through the regeneration system. In most plants, a flow controller regulates the volume of regeneration gas taken. This gas is sent through a heater, usually a salt-bath type, where it is heated to 400°F–450°F and then piped to the tower being regenerated.

The relationship between regeneration gas temperature and desiccant bed temperature for a typical 8-hour cycle is shown in figure 5.7. At about 240°F, water begins boiling, and the bed continues to heat up, but more slowly, since water is being driven out of the desiccant. After all the water has been removed, heating is continued to drive off any heavier hydrocarbons and contaminants, which will not vaporize at low temperatures. With cycle times of 4 hours or more, the bed will be properly regenerated when the outlet gas temperature has reached 350°F to 375°F. Following

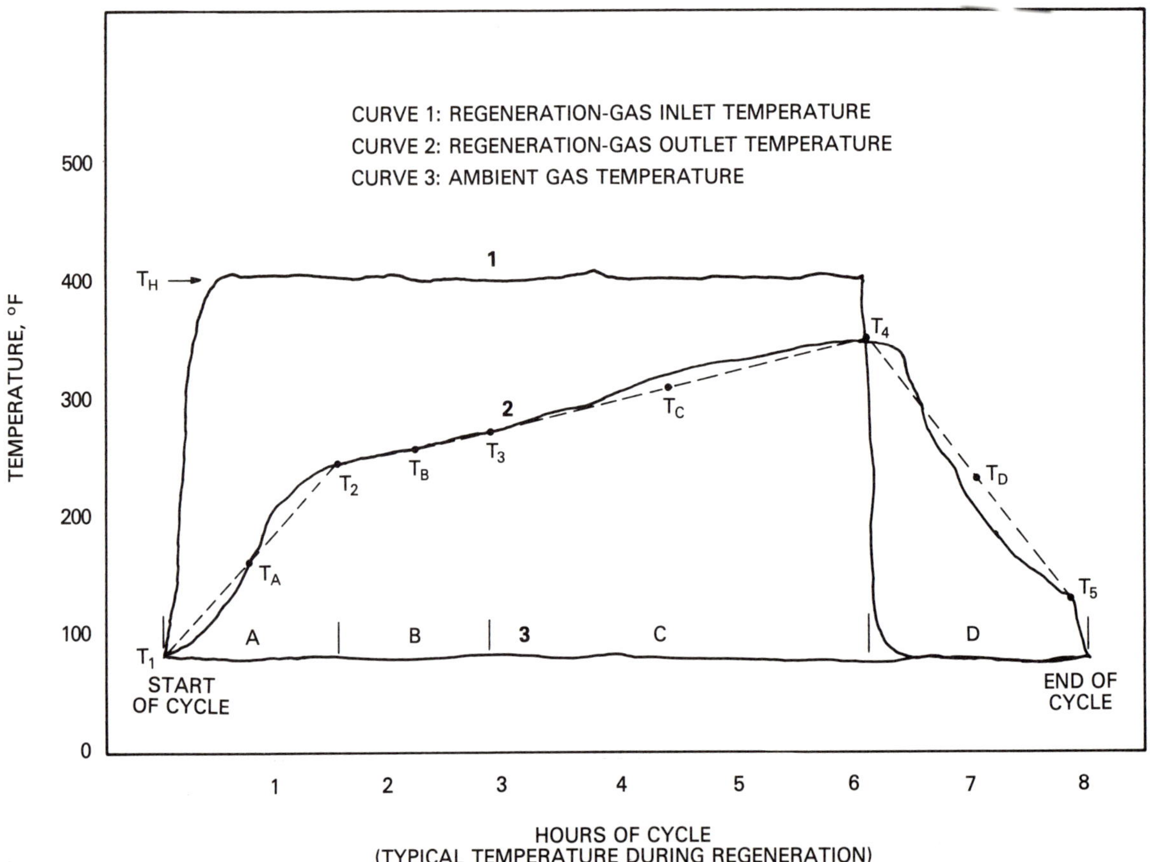

Figure 5.7. Regeneration-gas temperature versus desiccant-bed temperature in a dry-desiccant dehydrator

the heating cycle, the bed is cooled by flowing unheated regeneration gas through it. The cooling cycle is normally terminated when bed temperature has dropped to about 125°F, because further cooling may cause water to condense from the wet-gas stream and to presaturate the bed before the next adsorption cycle begins. All of the regeneration gas used in the heating and cooling cycle is passed through a heat exchanger, normally an aerial cooler, where it is cooled in order to condense the water removed from the regenerated tower. This water is separated in the regeneration gas separator, and the gas is then mixed with the incoming wet-gas stream. The entire procedure is continuous and automatic and basically a simple and trouble-free operation.

The usable life of a desiccant may range from 1 to 4 years in normal service. All desiccants become less effective in normal use through loss of effective surface area. This loss is rapid at first, but becomes more gradual as the desiccant ages. Abnormally fast degradation occurs with blockage of the small pores and capillary openings, which contain most of the effective surface area. Lubricating oils, amines, glycols, corrosion inhibitors, and other contaminants, which cannot be removed during the regeneration cycle, will eventually ruin the bed. Hydrogen sulfide may poison the desiccant and reduce its capacity. Light liquid hydrocarbons may accumulate if inadequate regeneration temperatures are not attained. Activated alumina has good resistance to liquids, but it may tend to powder, due to mechanical agitation of the flowing gas.

Automatically operated dry-desiccant dehydration units are very satisfactory for continuous operation. For units that have been well designed, most of the day-to-day operating problems are minor and of a mechanical nature. However, with long-time usage of the equipment and the desiccant, some serious operating problems may occur. The importance of protecting the desiccant beds from slugs of liquid water and liquid hydrocarbons should be emphasized. Some of the better desiccants now available are highly susceptible to failure when doused with liquid water.

Severe fouling of the dry desiccant bed may occur in a very short time when a gas stream containing both hydrogen sulfide and oxygen is treated. Although the relative concentrations of these two components may be very small, the desiccant acts as a catalyst to convert the hydrogen sulfide to free sulfur and water. Free sulfur in the desiccant plugs up the pores so that the desiccant has little capacity for removing water vapor.

Corrosion is usually not a serious problem in dry desiccant dehydration; however, when considerable carbon dioxide and hydrogen sulfide are present, corrosion may occur in the regeneration-gas heat exchanger. At this point, there is free water condensing inside the heat exchanger in the presence of these acid gases. Extreme corrosion has been noted where both carbon dioxide and oxygen are present in the gas to be dehydrated.

Occasionally corrosion in the regeneration equipment has been combated by the injection of ammonia ahead of the regeneration-gas cooler. Although this usually helps to control the pH of the water that is being removed, it has an adverse effect upon the operation of the dry desiccant. Ammonia and carbon dioxide react to form an unstable white solid that plugs up the pores of the desiccant. Upon termination of the injection of ammonia and repeated regeneration of the bed, much of the carbonate that has been formed will decompose, restoring some of the desiccant's capacity for removing water.

In some cases, continuous attrition of the desiccant has resulted in partial plugging of the bed and difficulty with compressors downstream. The desiccant breakage is usually the result of gas velocities being too high, slugs of water reaching the desiccant, or sudden pressure surges. It has frequently been necessary to install a special filter in the main-gas stream behind the desiccant beds.

Occasionally the main-gas temperature has been too hot immediately following the switch to the regenerated-desiccant bed. If this happens when the gas is flowing at the full design capacity of the unit, it is frequently the result of a heating cycle too long or a cooling cycle too short. Even if the outlet-gas temperature immediately after the tower switch is satisfactory at the full design capacity, this temperature becomes higher as the unit handles less gas. The condition results from the fact that the quantity of desiccant and the volume of steel necessary to be heated in the regeneration process is constant. As the quantity of main-gas flow decreases, there is less mass for cooling the reactivation-gas stream. The main-gas outlet temperature then must necessarily increase to dissipate the heat from the desiccant bed and the pressure vessel.

Most heaters for dry desiccant dehydration units utilize a high-temperature salt bath. Although somewhat more costly than a direct-fired tubular heater, it has several advantages. The high-temperature bath is kept at a constant temperature so that the heat is readily available for regeneration. Tube failure is greatly reduced by decreasing the metal wall temperature and eliminating severe hot spots. Direct-fired tubular heaters for this application are susceptible to failure. Such failure causes considerable damage, since the gas in the tubes is at full line pressure.

When a unit is operated at some pressure less than that for which the installation was designed, throughput should be reduced to maintain the desired efficiency. A given volume passing through the sorber at reduced pressure can do so only by reason of increased velocity. This, in turn, causes excessive pressure drop and may result in disturbance of the desiccant bed.

Performance is sensitive to the temperature of the gas supplied to the unit. All units are designed to handle a maximum volume of gas at a specified temperature and pressure. If the pressure remains constant and the inlet temperature rises, the throughput of the unit should be decreased.

Hydrocarbon recovery units

Methods and equipment for recovering the largest possible quantity of liquefied petroleum products from natural gas have undergone considerable change and improvement as the demand for LPG has increased. The development of hydrocarbon recovery units is an example. Although low-temperature separation units—employing either expanding gas or mechanical refrigeration for cooling the gas—and gas plants themselves are actually hydrocarbon recovery units, the term has come to apply primarily to dry desiccant adsorption units. These units can be designed to operate profitably on relatively small or quite large volumes (5 MMcf to 100 MMcf or more) of low-liquid gas—for example, 0.25 to 0.50 gallons of propane and heavier components per 1,000 cubic feet. The adsorption process employed in these units is a variation of the dry desiccant dehydration process.

In the dehydration process, water is adsorbed first in the top of the desiccant bed; and hydrocarbons, selectively heavier to lighter, are adsorbed in lower levels of the bed and are later replaced by water as the bed is saturated from top to bottom. For hydrocarbon recovery, the so-called quick-cycle system is used, in which the adsorption cycle is quite short, lasting only 15 to 20 minutes. This is long enough for the heavier hydrocarbons to replace the methane and ethane, but short enough so that the water does not replace the heavier hydrocarbons. Thus, instead of only water being driven off in the drying cycle, both water and hydrocarbons are driven off. These are condensed in coolers, the water being discarded and the hydrocarbons saved.

A flow diagram for a three-tower HRU open-cycle unit is shown in figure 5.8. The

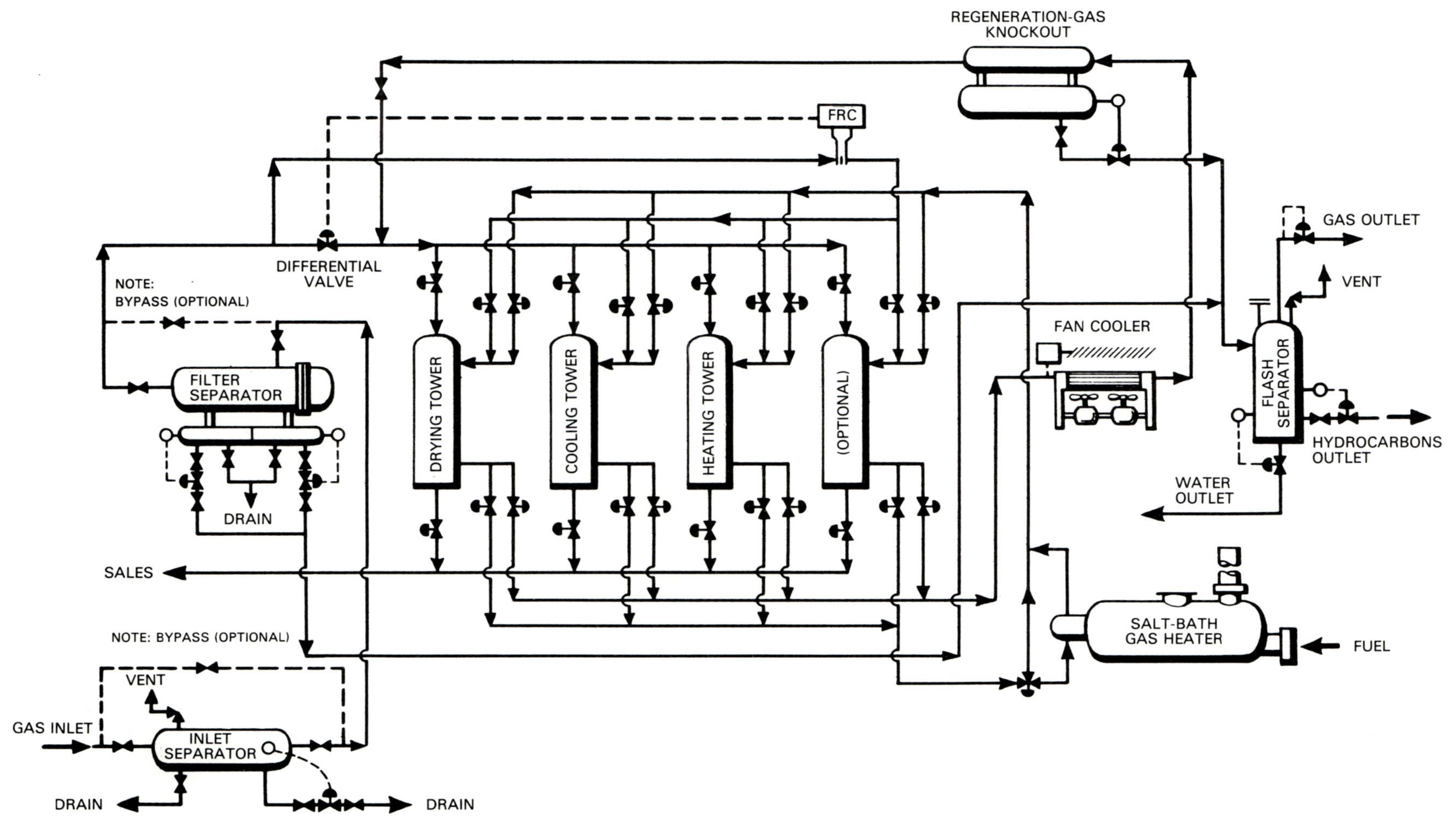

Figure 5.8. Flow diagram of three-tower HRU open cycle

equipment for the adsorption process is almost the same as that for dehydration, except that three or four contactors are needed to get adequate heating and cooling because of the short adsorption cycle. Normally, while one tower is in the adsorption cycle, one is cooling and one is heating or regenerating.

Advantages of the adsorption system over other hydrocarbon recovery systems are: (1) separate dehydration is not needed; (2) little moving machinery is required, making maintenance easier; and (3) high pressure is not necessary. In all systems, stage separation or stabilization is needed for the liquid production in order to hold evaporation losses to a minimum.

6 Miscellaneous gas conditioning

Natural gas well streams often contain hydrogen sulfide (H_2S) and carbon dioxide (CO_2). These two gases are called acid gases because in the presence of water they form acids or acidic solutions. These gases, particularly H_2S, are very undesirable contaminants, and unless they are present in very small quantities, they must be removed from a natural gas well stream.

Most pipeline specifications limit H_2S content to 0.25 grain per 100 cubic feet of gas. This is equivalent to about four parts per million. H_2S must be removed for several reasons, the most important being that it is a toxic and very poisonous gas and cannot be tolerated in gases that may be used for domestic fuels. H_2S in the presence of water is extremely corrosive and can cause premature failure of valves, pipelines, and pressure vessels. It can also cause catalyst poisoning in refinery vessels and requires many expensive precautionary measures to be taken.

The terms *sour crude* and *sour gas* refer to crude oil and gas that contain H_2S in amounts above acceptable industry limits. The terms *sweet crude* and *sweet gas* refer to oil and gas that do not contain H_2S or that have been treated to get rid of it.

Carbon dioxide removal is not always required, but most treating processes that remove H_2S will remove CO_2 also; therefore, the volume of CO_2 in the well stream must be added to the volume of H_2S to arrive at the total acid gas volume to be removed. CO_2 is corrosive in the presence of water, and as an inert gas has no heating value. In sufficient quantities, therefore, CO_2 might reduce the heating value (Btu/cf) below acceptable limits. Carbon dioxide removal may be required for gas going to cryogenic plants to prevent solidification of the CO_2.

Carbon disulfide (CS_2), carbonyl sulfide (COS), and mercaptans must also be considered in treating processes. Pipeline specifications normally allow 10 to 20 grains of total sulfur per 100 cubic feet of gas, including the 0.25 grain of H_2S. CS_2, COS, and mercaptans must be included in the total maximum allowable sulfur content.

Removal of acid gases

Several processes may be used to remove acid gases from natural gas. Some are selective for H_2S, others for CO_2. The oldest process, the iron sponge process, is also the most limited. It is a dry process using iron oxide (Fe_2O_3) impregnated on wood chips or shavings. It is usually used on low-concentration sour gas streams. The vessel can operate 30 to 60 days either without any regeneration or with the partial regeneration that can be effected with air passage through the vessel. The vessel must be recharged with new iron sponge material when gas sweetening is no longer possible.

The most widely used process in industry, the *alkanolamine process,* is a continuous-operation liquid process using absorption for the acid gas removal, with subsequent heat addition to strip the acid gas components from the absorbent solution. Most commonly used absorbing alkanolamine solutions are not selective and absorb total acid gas components. The process is particularly useful in order to obtain low acid gas residual concentrations, such as are required for gas transmission pipeline gas, chemical feedstocks, and domestic household and building heating usage. The type of process involved is chemical absorption; the absorbing alkanolamine solution reacts chemically with the absorbed components.

Iron sponge sweetening

The iron sponge process is one of the oldest known for removal of sulfur compounds from industrial gases, but only in recent years has the process been applied to gas sweetening of high-pressure natural gas. The process is a batch process, the sponge being a sensitive, hydrated iron oxide (Fe_2O_3) supported on wood shavings. The reaction between the sponge and the H_2S in the gas stream is—

$$2\ Fe_2O_3 + 6\ H_2S \rightarrow 2\ Fe_2S_3 + 6\ H_2O$$

The ferric oxide is present in a hydrated form, and without the water of hydration the reaction will not proceed. Thus, the operating temperature of the vessel must be kept below 120°F, or a supplemental water spray must be provided.

Regeneration of the bed is sometimes accomplished by the addition of air (O_2), either continuously or by batch addition. The regeneration reaction is—

$$2\ Fe_2S_3 + O_2 \rightarrow 2\ Fe_2O_3 + 6\ S$$

Because the sulfur remains in the bed, the number of regeneration steps is limited, and eventually the bed will have to be replaced.

The iron sponge process is most applicable for small gas volumes with low H_2S contents. It is not affected by pressure, and residue gas, well within the 0.25 grain H_2S/100 cu ft specification, can be obtained as long as the bed is not fouled. The process is selective toward H_2S. If there is CO_2 in the stream, it will not be affected. Also, for gas streams with small amounts of oxygen present, the O_2 serves to regenerate the bed continuously. The primary disadvantage of the process is the difficult changeout operation and disposal of the spent sponge.

Alkanolamine sweetening

The term *alkanolamine* encompasses the family of specific organic compounds including monoethanolamine (MEA), diethanolamine (DEA), and triethanolamine (TEA). These chemicals, and proprietary mixtures containing them and other amines, are used extensively for the removal of hydrogen sulfide and carbon dioxide from other gases and are particularly adapted for obtaining the low acid gas residuals that are usually specified by pipelines. Although some relatively new alkanolamines, such as methyldiethanolamine (MDEA), are selective to H_2S, the conventional amines (MEA and DEA) are applied in nonselective processes that remove both CO_2 and H_2S. If CO_2 removal is not required or desired to meet market specifications, then a solvent selective to H_2S should be considered.

The process is based on the chemical reaction of a weak base (alkanolamine) and a weak acid (H_2S and/or CO_2) to give a water-soluble salt. The following typifies the reactions of acid gases with monoethanolamine.

Absorbing:

$$\text{MEA} + H_2S \rightarrow \text{MEA hydrosulfide} + \text{heat}$$

$$\text{MEA} + H_2O + CO_2 \rightarrow \text{MEA carbonate} + \text{heat}$$

Regenerating:

$$\text{MEA hydrosulfide} + \text{heat} \rightarrow \text{MEA} + H_2S$$

$$\text{MEA carbonate} + \text{heat} \rightarrow \text{MEA} + H_2O + CO_2$$

These are reversible equilibrium reactions with an increase in temperature shifting the equilibrium to the left.

In general, most plants being built today for operation above 125 psia utilize DEA because lower circulation rates are required and less energy is used for regeneration (fig. 6.1). At pressures below 50 psia, DEA will not reduce the total sulfur content sufficiently to achieve the 0.25 gr/100 cf usually required by pipeline specifications. MEA is preferred for these low-pressure applications.

Figure 6.1. DEA desulfurizer for offshore sour gas (Courtesy of Black Sivalls & Bryson)

MEA has the disadvantage of combining with carbonyl sulfide to form a nonregenerable compound and expending amine in this reaction. With streams containing appreciable amounts of COS, DEA is normally used, since its reaction product can be regenerated. COS is usually found in refinery cracked-gas streams but is not normally found in natural gases. However, some extremely sour natural gases do contain significant amounts of carbonyl sulfide.

The schematic flow diagram, figure 6.2, shows the basic process. The gas to be sweetened flows from its source, such as

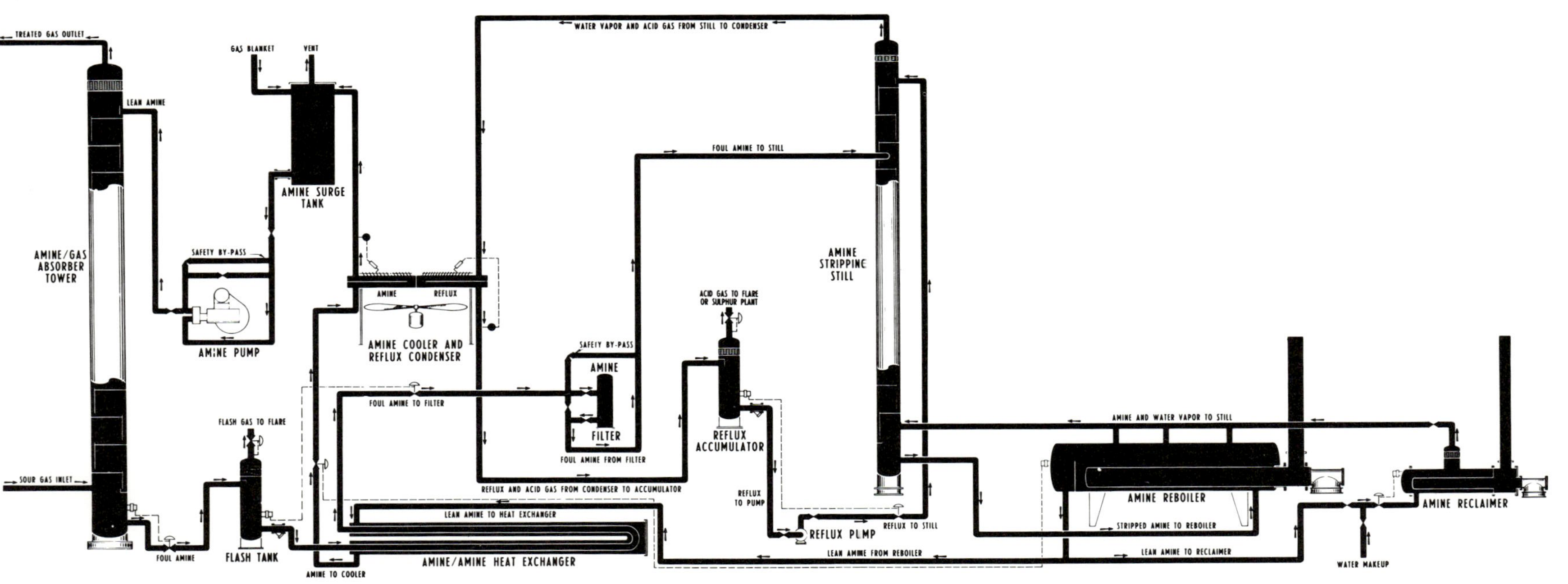

Figure 6.2. Schematic flow diagram of the alkanolamine sweetening process

wells, compressor discharge, and so forth, to the inlet scrubber. This scrubber can be an integral portion of the absorber, or it can be a separate vessel, dependent upon preference and operating conditions. It is extremely important that any separable liquids be removed from the gas stream prior to the absorption process.

After removal of any separable water or hydrocarbons, the gas passes to the absorber section and rises countercurrently in intimate contact with the descending amine solution. Purified gas flows from the top of the absorber to the processing plant, the dehydration unit, compression, sales, or other disposition. A scrubber on the absorber outlet is very desirable to catch any amine solution carryover resulting from periods of upset or unstable operation. Any amine caught in the outlet scrubber is automatically dumped back into the system via the amine surge tank.

The amine solution that circulates through the system contains from 10% to 65% amine in water, depending on the type of amine, the required amount of acid gas to be removed, and corrosivity. The lean amine solution flows across the top tray and down across subsequent trays and removes acid gases from the rising gas stream. The resulting amine rich with acid gas then leaves the bottom of the absorber through a dump valve actuated by a liquid-level controller. The rich amine then goes to the flash tank, operating at a reduced pressure, where a great portion of the physically absorbed gases are flashed off.

The rich amine then flows through the heat exchanger, where it is heated by the hot lean solution coming from the still, and its temperature is increased to approximately 190°F. Then it flows into the top of the regeneration column and is further heated by steam rising countercurrently to the descending amine solution. This heating shifts the equilibrium, which liberates H_2S and CO_2 and regenerates the MEA. The steam also serves to sweep the liberated acid gases out of the still.

Overhead vapors from the still, consisting essentially of steam and acid gases, go to the reflux condenser, where the steam is condensed and the acid gases cooled. This cooling can be accomplished with water, with air, or with the gas stream in some special cases.

The condenser effluent then flows to the reflux accumulator for separation of the condensed steam and acid gases. The steam condensate is returned to the still by the reflux pump. Acid gases are removed from the system through the still back-pressure valve and may go to a flare stack or a sulfur-extraction facility.

Hot regenerated amine from the still kettle flows through the heat exchanger, where it is cooled by the rich amine stream. Final cooling is accomplished by aerial coolers or by water coolers, if available. It then flows to the surge tank, with flow controlled by a liquid-level controller and diaphragm valve to maintain a constant level in the reboiler. Amine pumps take suction from the surge tank and discharge to the top tray of the absorber to complete the cycle.

Heat is added to the still reboiler by steam, hot oil, or direct firing. To reduce water losses, the steam system may be a closed-circuit integrated facility with gravity return of condensate to the steam generator. It may also be the existing plant steam for turbines, boilers, and so forth. Hot oil heating is accomplished by heating an oil-transfer medium in a direct-fired heater and pumping the oil through a reboiler tube bundle. The direct-fired method of heating consists simply of controlled combustion on the inside of an expandable tube that is surrounded by the boiling amine solution.

Glycol/amine process

The glycol/amine process uses a solution comprised of 10 to 30 weight percent MEA, 45 to 85 percent glycol, and 5 to 25 percent water for the simultaneous removal of water vapor, H_2S, and CO_2 from gas streams. The combination dehydration and sweetening unit results in lower equipment cost than would be required with the standard MEA unit followed by a separate glycol dehydrator. Obtaining pipeline specification for H_2S in the residue gas stream does not appear to be a problem with glycol/amine, since the solution, with the high boiling point glycol present, can be stripped at atmospheric pressure and fairly high temperatures. The same is not true in conventional MEA units, since the high temperatures needed for more complete stripping adversely affect the MEA solution and lead to excessive degradation and corrosion problems. The degree of dehydration that can be obtained from the glycol/amine process will be as good as or slightly better than that from a glycol solution with an equal amount of water, that is, a 95 percent DEG and 5 percent H_2O mixture. This should not be confused with the dehydration that can be obtained from a standard glycol dehydrator, which produces glycol with purities in excess of 99 percent.

The main disadvantages of the glycol/amine process are as follows:

1. Increased vaporization losses of MEA are due to the higher regeneration temperature.
2. Reclaiming must be by vacuum distillation.
3. Intricate corrosion problems are present in operating units, and the solution of one plant may not apply to any other plant.
4. Application must be for gas streams that do not require low dew points.

The process flow scheme is essentially the same as that for MEA.

Sulfinol process

The Shell sulfinol process is unique in that it uses a mixture of solvents, allowing it to behave as both a chemical and a physical solvent process. The solvent is composed of sulfolane, DIPA (di-isopropanolamine), and water. The relative amounts of each ingredient are varied to yield a solvent composition that is tailored to fit each application. The sulfolane acts as the physical solvent, whereas DIPA acts as the chemical solvent. This combination of absorption capabilities offers advantages both for loading and for unloading the solvent. The sulfinol solvent has a good affinity for sour components at low to medium partial pressures and an extremely high affinity for sour components at high partial pressures.

Sulfinol appears to have its greatest advantage when the acid gas partial pressure is about 30 psia or greater. In general, COS, CS_2, and mercaptans can be satisfactorily removed from the feed gas, along with H_2S and CO_2, within certain limitations of concentrations in the feed gas. Degradation of sulfinol solvent by COS, CS_2, and H_2S is practically nil. CO_2 gradually degrades the DIPA to DIPA-oxazolidone, which can be readily removed from the system by a simple reclaimer unit.

The main advantages of sulfinol are (1) low solvent circulation rates; (2) smaller plant equipment and lower plant cost; (3) low heat capacity of the solvent; (4) low utility costs; (5) low degradation rates; (6) low corrosion rates; (7) low foaming tendency; (8) high effectiveness for removal of COS, CS_2, and mercaptans; (9) low vaporization losses of the solvent; (10) low heat-exchanger fouling tendency; and (11) nonexpansion of the solvent when it freezes.

Disadvantages of sulfinol are (1) the absorption of heavy hydrocarbons and aromatics; and (2) expense of the sulfolane in the solvent. Sulfinol is a proprietary process and requires payment of a royalty to Shell for its use.

Other proprietary solvents have been developed for gas sweetening. The selection of which solvent to use for a certain application depends on such factors as extent of H_2S and CO_2 removal, gas composition (H_2S, CO_2, COS, and heavier hydrocarbons), and gas pressure and temperature.

Molecular sieve removal of H_2S and CO_2

In addition to the amine-solution and sulfinol-solution processes, H_2S and CO_2 can be removed by the use of mol sieves in a dry bed unit such as that discussed previously. Today the major difficulty with these units is how to dispose of the sour regeneration gas. Usually this gas is flared or burned in an incinerator. With the emphasis on environmental protection today, operators are finding it more and more difficult to vent or burn any plant effluents such as these. Some consideration has been given to using a small MEA system in combination with a sulfur plant to handle these effluents.

Btu control

As more and more ethane and propane are extracted from natural gas, the control of the Btu, or heating content, of the gas becomes more important. Small, low-temperature gasoline plants are being installed, capable of removing all of the propane and heavier components and 60–70 percent of the ethane. The residue gas from such plants often consists of little more than methane plus inert gases such as CO_2 or N_2. If the feed gas contains substantial quantities of such gases, the heating value of the residue gas may fall below contract specifications (usually in the order of 1,000 Btu/cf). Thus, the amount of ethane or propane extracted may be limited to something less than that economically possible otherwise.

When possible, low-Btu gases are blended with gases having fairly high Btu values so that the average meets specifications. Such blending requires a high degree of care and precision, especially if the gas is to be used for residential heating. Variations in the gas composition will cause smoking or flameouts because the air-mixture setting on burners is fixed. In most cases where blending takes place or plant operations can cause a variation in the Btu content, recording calorimeters are used to monitor the residue-gas stream. Adjustments are made in the blending or plant operation to maintain constant a Btu content based on data obtained from the calorimeter.

7 Compressors and prime movers

A most common item of equipment used in the handling and transporting of gas is the compressor. Compressors may vary in size from small belt-driven units in the order of 50 hp to very large reciprocating or centrifugal units of 15,000 hp or more. The most common type of compressor is the gas-engine-driven reciprocating unit. However, gas-turbine-driven centrifugal units are coming into use more and more. Other types of drivers for reciprocating units are electric motors and steam turbines. These drivers are ordinarily used in special situations and are rarely found in field operations. Electric motors are normally difficult to justify, since electric power usually costs more than the fuel gas that would be saved. However, electric motors are being used in more and more applications where the exhaust emissions from gas engines or turbines would exceed environmental limits. Expansion turbines are used quite often in plants or refineries where a pressure drop in a fluid stream (gas, hot oil, etc.) can be utilized.

Reciprocating compressors

Drivers

The internal-combustion gas engine is the leading prime mover for gas compression service in the oil industry. The popularity of the gas engine in this field has been maintained primarily because of its overall efficiency, because of the availability of a clean, relatively low-cost fuel, and because manufacturers of this equipment have made gas engines and reciprocating compressors into compact integral units. Horsepower ratings and general design of modern gas engines are significantly different from the units manufactured a generation ago. Today, integral gas engine-compressor units are available from under 200 hp to 7,500 hp or more. Guaranteed thermal efficiencies of some of the gas engine-compressor units are now approaching 40 percent. For comparison, maximum thermal efficiency of a steam turbine is approximately 33 percent. Gas-turbine-engine thermal efficiency may be less than 25 percent, depending to a great extent on the utilization of the heat in the exhaust gases.

Gas engines may be classified in a number of ways including the following.

1. Combustion cycle
    - a. Four-stroke cycle
    - b. Two-stroke cycle
2. Power impulse
    - a. Single acting
    - b. Double acting
3. Cylinder arrangement
    - a. Vertical
    - b. Horizontal
    - c. V-type
    - d. Opposed
    - e. Radial
4. Speed
    - a. Low (100 to 250 rpm)
    - b. Intermediate (250 to 600 rpm)
    - c. High (600 to 1,200 rpm)

In general, the gas engines most commonly used now are both two- and four-stroke cycle units with single-acting power cylinders in a vertical or V-type arrangement, operating in the intermediate or high speed range. The smaller units connected to compressors by

Figure 7.1. Gas engine connected to compressor by belt

belts (fig. 7.1) are all considered to be high-speed units. Direct-drive units (fig. 7.2), those in which the engine crankshaft is connected to the compressor crankshaft with a coupling or clutch, usually operate at high speeds. Integral units (fig. 7.3), those that provide for mounting of compressor cylinders on the engine frame, operate at low to intermediate speeds. All of the various types of gas engines are basically reciprocating machines with rotating crankshafts; their fundamental difference lies in the combustion cycle. Neither four- nor two-stroke types have any clear-cut overall advantage; years of competition among engine builders have failed to eliminate either type.

The conventional two-stroke cycle engine consists of a combination gas engine and scavenging cylinder (air compressor) built into one compact unit (fig. 7.4*A*). Two-stroke cycle gas engines require two piston strokes and one full revolution for each cycle. The exhaust ports in the cylinder walls are uncovered by the piston near the end of the expansion stroke, permitting the escape of exhaust gases and reducing the pressure in the cylinder. The charge of air flows into a scavenger cylinder and is compressed to a few pounds per square inch above atmospheric pressure. Intake ports are uncovered by the piston soon after the opening of the exhaust ports, and the compressed charge of air flows into the cylinder expelling most of the exhaust

Figure 7.2. Direct-drive reciprocating gas compressor (Courtesy of Ingersoll-Rand)

Figure 7.3. Integrally connected prime mover and compressor unit

products (fig. 7.4*B*). When the piston starts toward the cylinder head, it closes off the intake ports and then the exhaust ports. The fuel gas is injected into the cylinder through a valve in the cylinder head and is mixed with the supply of air that has just been admitted. As the piston moves toward the cylinder head, the compression of the fuel gas and air mixture continues until ignition, after which the piston is forced back as the charge expands.

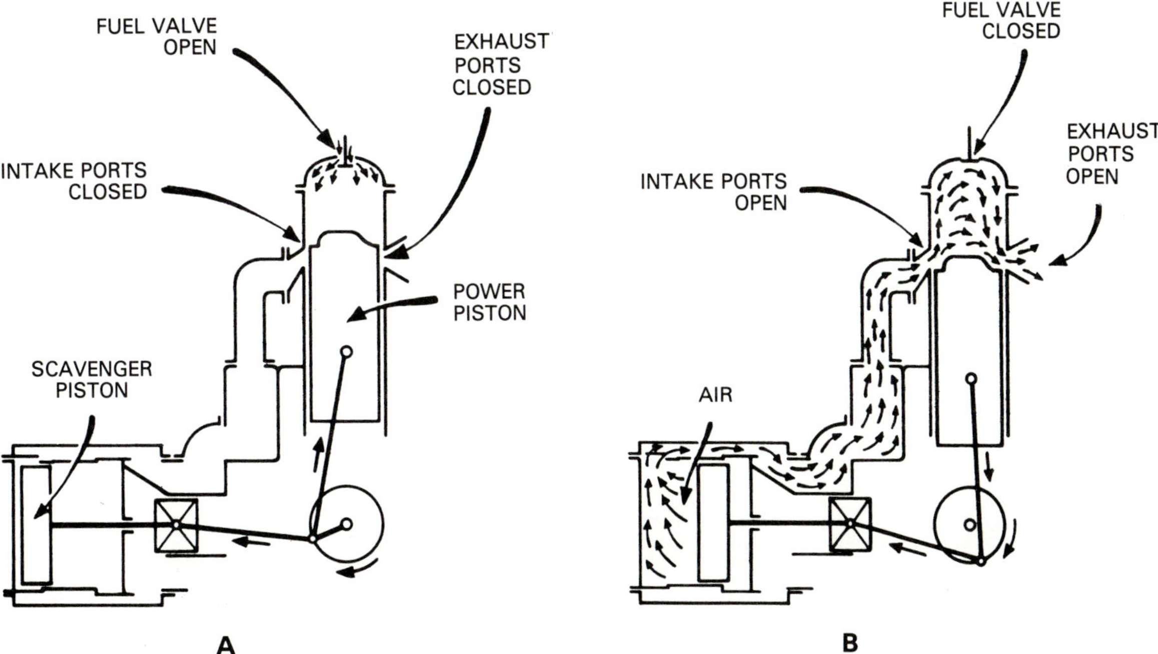

Figure 7.4. Schematic sketch of a two-stroke cycle engine operation

As the piston approaches the end of its power stroke, the exhaust ports are uncovered, as mentioned above, completing the cycle. Construction of a typical two-cycle engine is shown in figure 7.5.

In the operation of a conventional four-cycle engine (fig. 7.6), four piston strokes and two complete revolutions are required to complete each cycle. Both intake and exhaust valves are closed when the piston is near the end of its compression stroke, and ignition occurs. The explosion forces the piston away from the cylinder head. When the power stroke is ended, the piston moves toward the cylinder head, the exhaust valves open, and the exhaust products are expelled. As the piston starts toward the crankshaft for the second time, the intake valve opens, and the fuel-air mixture is drawn into the cylinder. The piston then starts toward the cylinder head again on the compression stroke. Ignition occurs when the piston nears the head end of the cylinder, which is at the top center of the crankshaft revolution, and the cycle is completed. It will be noted that four strokes are required to complete this cycle: (1) intake, (2) compression, (3) power, and (4) exhaust. Construction of a typical four-cycle engine is shown in figure 7.7.

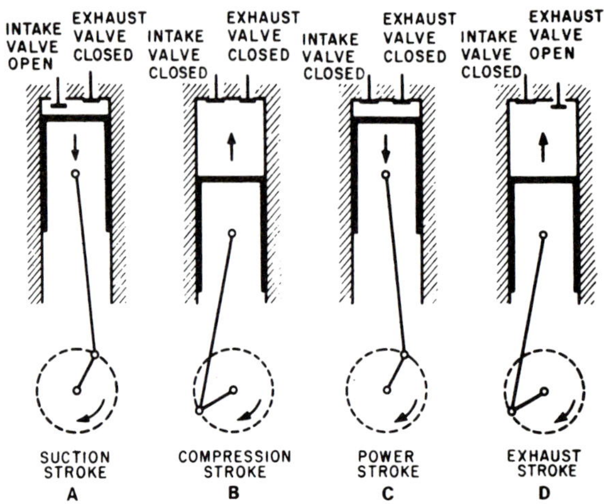

Figure 7.6. Schematic sketch of a four-stroke cycle engine operation

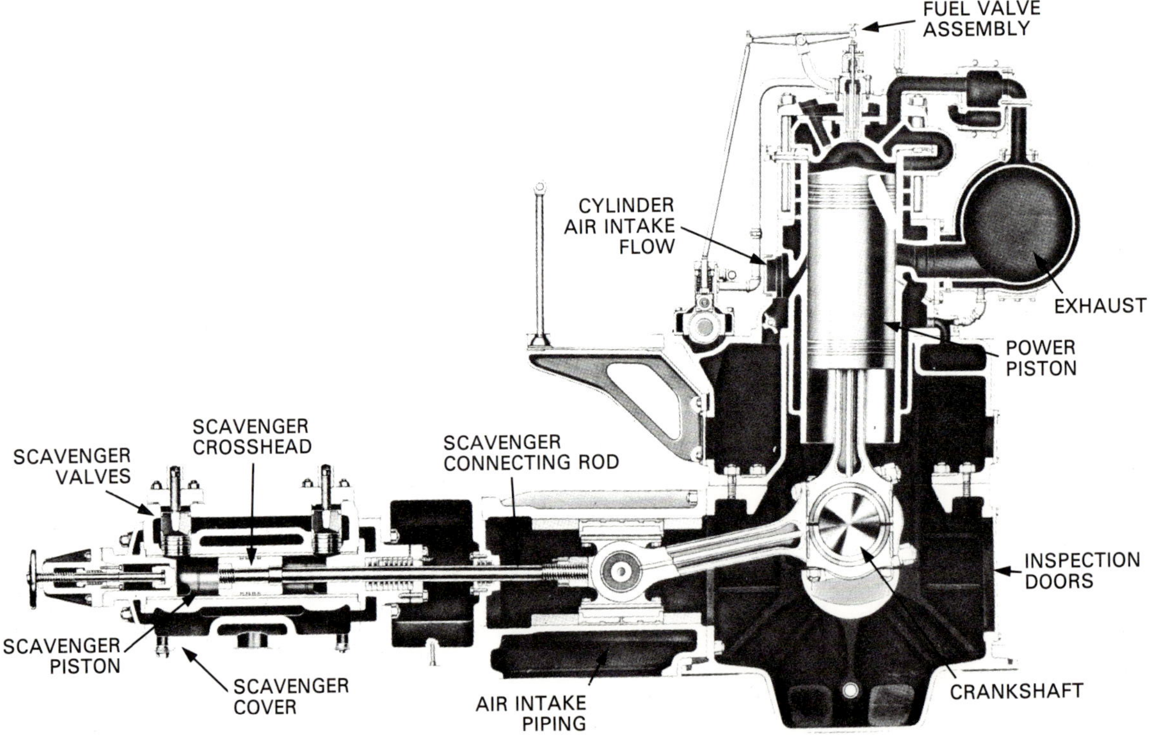

Figure 7.5. Cross section of a two-stroke cycle gas-engine compressor

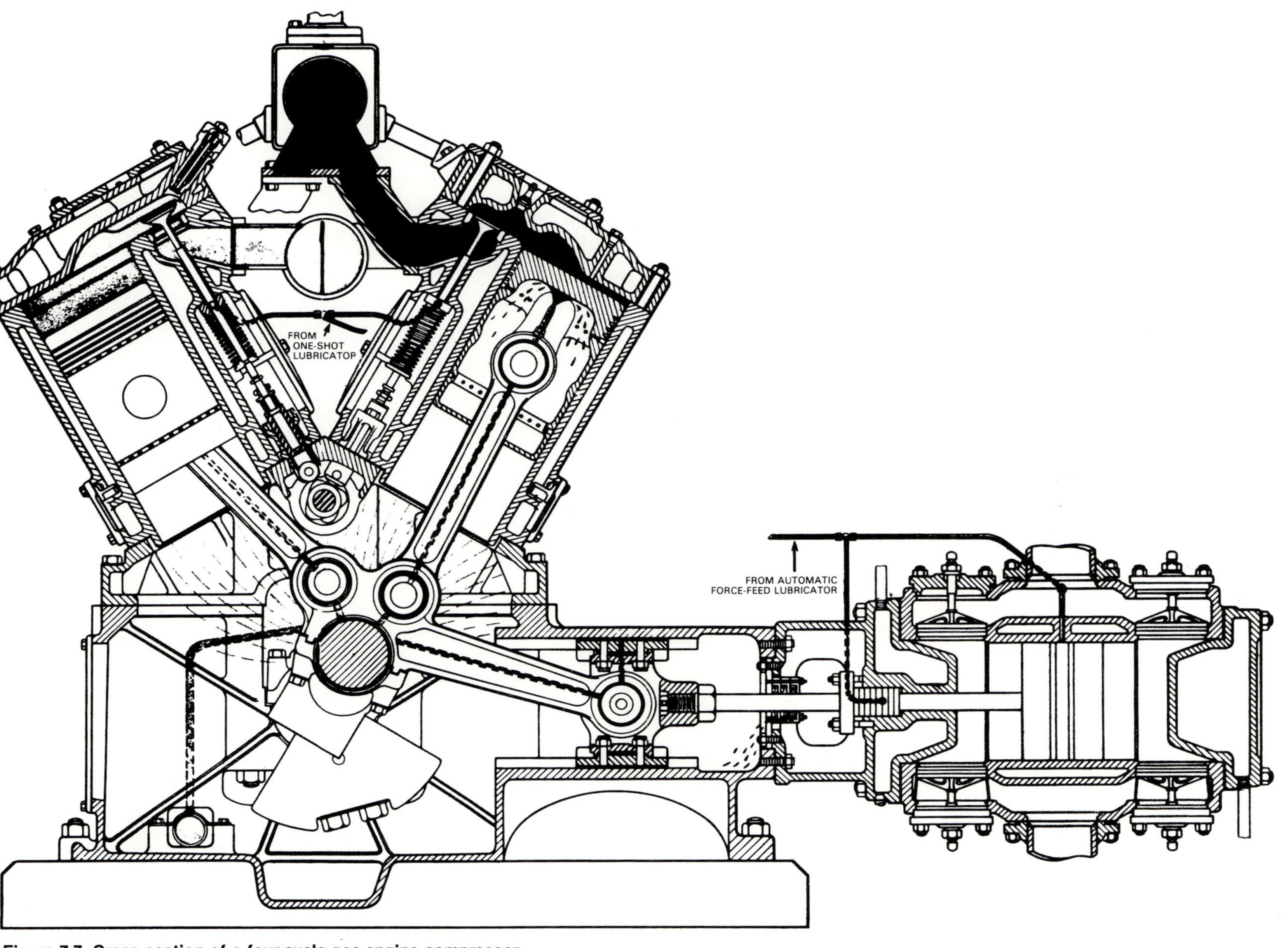

Figure 7.7. Cross section of a four-cycle gas-engine compressor

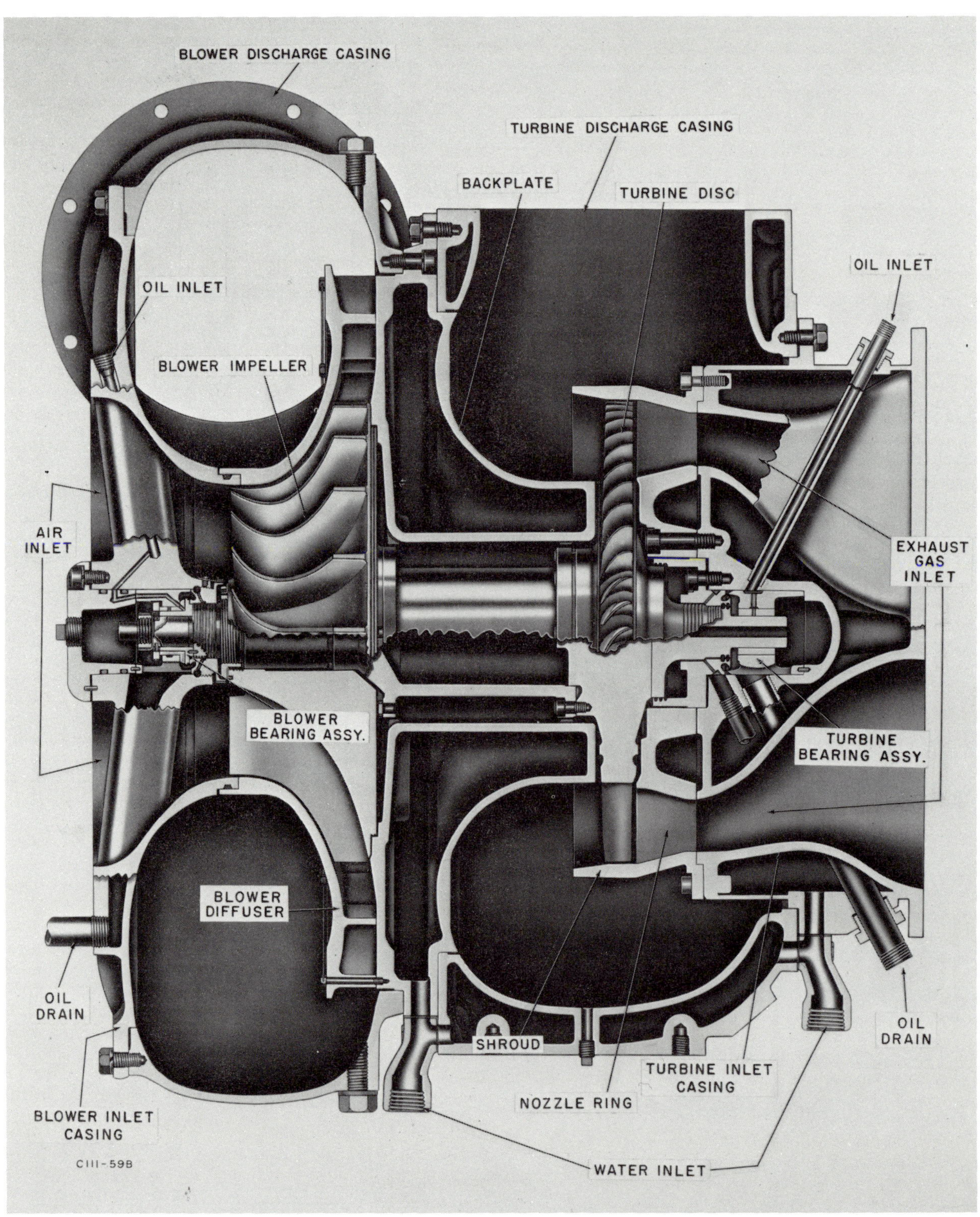

Figure 7.8. Cross section of an exhaust-gas turbine-driven centrifugal compressor for supercharging

The following generalizations can be made about two- and four-stroke cycle gas engines.

1. The fuel consumption of slow-speed integral-type units varies between 9,000 and 7,000 Btu/bhp/h depending upon the type, size, and make, with the lower fuel consumption resulting from turbocharging.
2. Maintenance costs are approximately the same for the two types. Generally speaking, however, as the size of the engine increases, the unit costs in terms of dollars per brake horsepower per year ($/bhp/yr) will gradually decrease.

Horsepower ratings of engines depend mainly upon the amount of air that can be supplied to the power cylinders. Because a very definite air-fuel ratio is required, an important factor in the design of engines is the method of supplying the air for combustion. In naturally aspirated four-stroke cycle engines, the vacuum created by the piston moving downward sucks air into the cylinder when the intake valve is open. Conventional two-stroke cycle engines, of course, have reciprocating air-compressor cylinders built integrally into the unit to provide the necessary air for scavenging and combustion. In recent years, attention has focused on ways and means of providing additional air so that additional fuel may be burned in order that the horsepower rating of an engine can be increased without altering the engine frame. The injection of additional air into the power cylinders so more fuel can be used is called *supercharging.*

Supercharging an engine may be done in several ways.

1. Using additional air-scavenging cylinders, which is possible only if space for an additional cylinder is available and only for two-stroke cycle engines
2. Using larger air-scavenging cylinders, which is possible only for two-stroke cycle engines
3. Using air blowers driven by belts off the engine flywheel
4. Using an electrically driven blower
5. Using a centrifugal compressor driven by an exhaust gas turbine, called a turbocharger

All of the above methods have been tried successfully by various companies. Turbocharging is a common method of increasing the power output and fuel efficiency of new gas engines (fig. 7.8). Supercharging of old units is possible but should be done only after consultation with the manufacturer to determine that the bearings, frame, and crankshaft can satisfactorily handle the increased mechanical load.

Compressors

The vast majority of compressors purchased for use in gas operations are the reciprocating piston type. Basically this type of compressor consists of a ringed piston, cylinder, cylinder head, and suction and discharge valves, plus the mechanism necessary to convert rotary motion to reciprocating motion—that is, connecting rod, crosshead, wrist pin, and piston rod (fig. 7.9). Valves are of the automatic plate type, which depend on differential pressure for opening and closing.

The compressor cylinder itself is fabricated of cast iron, nodular iron, cast steel, or forged steel. These materials are chosen in order to achieve successively higher maximum working pressures. Depending to a large extent on cylinder size, the cast-iron cylinders, which are generally the most economical, are satisfactory for working pressures as high as 1,500 psi. In the larger sizes, of course, the working pressure of cast-iron cylinders must be reduced drastically, and in such cases nodular iron may be substituted to increase the working pressure. Nodular iron, sometimes called spheroidal or ductile iron, is in reality a high-grade cast iron in which the graphite present has been arranged in substantially spherical shape rather than in the usual

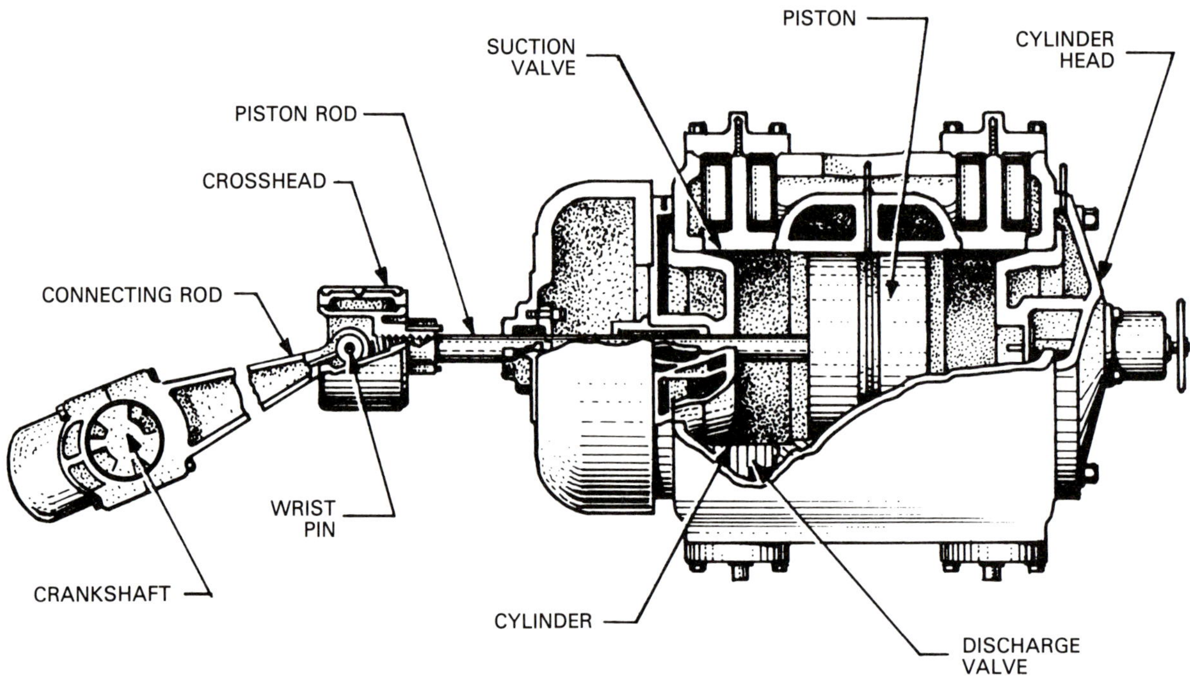

Figure 7.9. Elements of a typical reciprocating compressor

flake form, as in gray cast iron. The relative tensile strengths are roughly 45,000 to 67,000 psi. Cast-steel cylinders are usually rated to working pressures of 2,500 to 3,000 psi, and forged-steel cylinders are used for higher pressures.

Reciprocating compressor cylinders may be purchased in either solid body or liner type. Solid-body cylinders are limited to one diameter and can be altered in size only by boring out the cylinder. Such altering can be done only within narrow limits because the cylinder working pressure is reduced by the change. Liner cylinders are more flexible because a machined liner is installed inside the body of the cylinder to provide the desired bore. Should the need for a different-size cylinder occur, the liner can be removed and a different-size liner and piston fitted into the body of the cylinder. Naturally, there are limitations as to the magnitude of diameter variation in each direction.

Reciprocating compressors are adaptable to low or high ratios and varying rates and pressure conditions; they can be built to handle small, intermediate, or high throughput volumes. They are therefore eminently suitable to gas production operations.

Reciprocating compressor operations

Theoretical considerations

Gas laws are commonly considered to be laws of nature that govern the behavior of gases under specified conditions. As discussed previously, gases react to pressure and temperature according to well-defined rules. These laws are directly applicable to the handling and processing of natural gas.

Work is performed when a force acts through a distance or the resistance to motion is overcome. No work is done unless motion is produced. Although gas molecules are constantly in motion overcoming internal resistance and causing pressure in a closed container, work is performed only when this pressure or force causes movement through expansion, such as pushing on a moving piston. *Force* is that which causes, changes, or stops the motion of a body. Mathematically, work is force times distance.

Energy is the capability of a body for doing work. Potential energy is this capability due to the position or state of the body. Kinetic energy is the capability due to the motion of the body. A pendulum, swinging back and forth, constantly changing its motion is simply a trading of potential and kinetic energy. Gas molecules in motion are doing the same thing that the pendulum is doing. The amount of work actually done depends upon the amount of energy available. If an engine is capable of using more gas for fuel and thus doing more work, it is capable of just that much more energy. Molecules of gas may be moving at a certain rate and thus producing a certain pressure, but if an increase in temperature will cause them to move twice as rapidly, then their energy and resulting pressure are increased.

Power is the rate of doing work, or the amount of work done in a specific unit of time. It is calculated in foot-pounds per minute. One horsepower equals 33,000 foot-pounds per minute.

Velocity is the speed of a body in motion and is measured in distance per unit of time, such as feet per second. Gas molecules are minute bodies of matter that are always in motion except at absolute zero temperature. Their velocity determines the pressure a gas exerts and has a bearing on the work required to process, compress, or cool the gas.

Isothermal compression of air or gas exists when the interchange of heat between the air or gas and surrounding bodies (i.e., cylinders or pistons) takes place at a rate exactly sufficient to maintain the air or gas at constant temperature as the pressure increases. *Adiabatic compression* of air or gas exists when no heat is transferred between the air or gas and surrounding bodies (such as the cylinders and pistons in a compressor). It is characterized by an increase in temperature during compression and a decrease in temperature during expansion. No compressor operates under either condition perfectly, but its operation may be between the two. Some conditions may make one compressor operate near the isothermal and another near the adiabatic compression cycle.

Specific heat, expressed in Btu's, is the amount of heat required to raise 1 pound of gas 1 degree Fahrenheit. The amount varies, depending upon whether pressure or volume is held constant. The specific heat at constant pressure is denoted by C_p. The specific heat at constant volume is denoted by C_v. The specific heat will also vary with the temperature for all except the monatomic gases, which have only one atom in each molecule (e.g., helium, neon, krypton, xenon, and mercury vapor). The specific heat of other gases, such as oxygen, hydrogen, nitrogen, carbon dioxide, steam, and methane, are dependent upon the temperature range through which calculations are to be made. The operator must depend upon tables of collected data and calculations from actual compressor operation to determine their behavior. The ratio between the specific heat at constant pressure and the specific heat at constant volume is known as the *N* or *K* value of the gas. For example, it has been experimentally found that it takes 0.2375 Btu to raise 1 lb of dry air at 60°F, 1°F at constant pressure. If the volume had remained constant and the pressure had been allowed to increase, it would have required 0.1689 Btu. The *N* value in this case would be—

$$N = \frac{C_p}{C_v} = \frac{0.2375}{0.1689} = 1.406$$

The *N* value depends upon the specific property of the gas. The *N* value for air is 1.406, approximately 1.265 for dry natural gas, and 1.1 for wet casinghead gas. Since there is such a wide variation in wetness and *N* value of natural gas, a curve has been prepared that gives the N value as a function of the molecular weight or specific gravity of the particular gas. If a fractional analysis is available for the gas, the molecular weight may be calculated and used to determine the N value. If the molecular weight is not available, the specific gravity can be determined and used. The power required to compress a given volume of gas is affected appreciably by the N value of the gas.

Compression ratio, clearance, and volumetric efficiency

The *ratio of compression* of a compressor is the ratio of the absolute discharge pressure to the absolute suction pressure. For instance, a compressor with atmospheric intake at 0 psig (14.7 psia) and discharge at 40 psig (54.7 psia) has a ratio of compression of 54.7 to 14.7; that is, 3.72 to 1. This must not be confused with the compression ratio used in power cylinders of internal-combustion engines, which is the displacement plus the clearance divided by the clearance.

The ratio of compression for gas compressors is generally limited to less than 5.5, usually falling between 2.5 and 5.0. The reason is evident if the temperature and volumetric efficiency of the compressor are considered. When compression is necessary over a larger number of compressions than this, two or more stages of compression are used. For illustration, the following example is given. Assume that the discharge pressure is 4,000 psig and the suction pressure is 500 psig. The overall compression ratio is:

$$\frac{4000 + 14.7}{500 + 14.7} = \frac{4014.7}{514.7} = 7.80:1$$

Inasmuch as it is impractical to compress through 7.80 ratios in a single stage because of the heat of compression, two-stage compression will be required, and the ratio per stage will be the square root of 7.80, or 2.79. A comparable situation is possible with low pressures, that is, suction pressure of 15 psig with a discharge of 225 psig.

The volume delivered, regardless of the compressor stroke, will be affected by the *clearance* of the unit. As the piston moves back and forth within its cylinder, it must never reach the end of the cylinder or it would damage itself and the cylinder head. The space provided for this protection plus that which exists around the valves, which are located near the cylinder ends, make up the volume known as the clearance. The clearance volume is usually expressed as a percent of piston displacement volume by using the following formula:

$$\text{Percent Clearance} = \frac{\text{clearance volume, cu in.}}{\text{piston displacement volume, cu in.}} \times 100$$

Percent clearance for double-acting cylinders is based on total clearance volume in cubic inches for both head and crank end and total piston displacement volume for both head and crank end. Often the clearance in cubic inches is assumed to be equal between the head and the crank ends even though the crank-end displacement is less than the head-end displacement because of the rod displacement volume. For small-diameter cylinders, this difference in clearance volume can be appreciable and must be taken into account in calculations.

Volumetric efficiency, E_v, represents the efficiency of a compressor cylinder in compressing gas. It may be defined as the ratio of the volume of gas actually delivered, corrected to suction temperature and pressure, to the piston displacement volume. The principal reasons that the cylinder will not deliver the piston displacement capacity are: (1) wire-drawing, which is a throttling effect at the valves; (2) heating of the gas during admission to the cylinder; (3) leakage past valves and piston rings; and (4) reexpansion of the gas trapped in the clearance-volume space from the previous stroke. Reexpansion has by far the greatest effect on volumetric efficiency. Without going into the details of the derivation of the volumetric efficiency, which may be obtained in any thermodynamics text, the percent volumetric efficiency for natural gas may be calculated using one of the following equations:

$$E_v = \frac{100 - R}{100} - Cl\,(R^{1/N} - 1)$$

or

$$E_v = 0.97 - Cl\,[(R^{1/N} \times Z_s/Z_d) - 1]$$

where

E_v = volumetric efficiency, expressed as a decimal (for example, 0.8)
R = compression ratio
Cl = clearance, expressed as a decimal
N = ratio of specific heats
Z_s = compressibility factor at suction P and T
Z_d = compressibility factor at discharge P and T.

The terms $(100 - R)/100$ and 0.97 will obviously be equal at a compression ratio of 3.0. This factor in the equation provides for the losses due to wiredrawing, heating of the gas, and leakage, which are the first three items mentioned above. The two equations are given because there is a disagreement between manufacturers as to the proper approach to be used, since the effect of these three items cannot be measured. The volumetric efficiency obtained by either will be approximately the same over the range of compression ratios normally used in gas operations.

Cylinder capacities

Solving compressor problems can be reduced to a relatively simple sequence of calculations by applying a few basic equations and utilizing data taken from curves contained in the GPSA *Engineering Data Book*. However, it is always desirable to know about the origin of the basic terms, and for this reason, the section below has been included. Figure 7.10 shows a typical ideal pressure-volume diagram for a compressor cylinder with corresponding compressor piston locations during reciprocation.

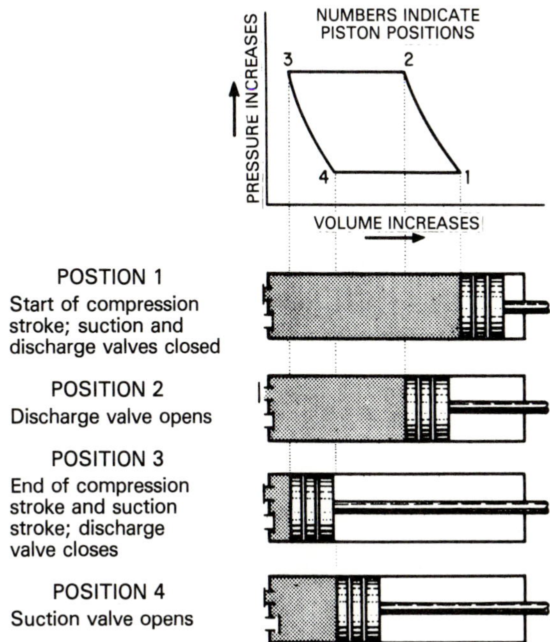

Figure 7.10. Pressure-volume diagram and piston locations during a reciprocating compressor stroke

Position 1. This is the start of the compression stroke. The cylinder has a full charge of gas at suction pressure. As the piston moves toward position 2, the gas is compressed along the line 1 to 2.

Position 2. At this point, the pressure in the cylinder has become greater than the pressure in the discharge line, causing the discharge valve to open and allowing the original charge of gas to enter the discharge line. This action occurs along the line 2 to 3.

Position 3. At this point, the piston has completed its discharge stroke, and as soon as it starts its return stroke, the pressure in the cylinder drops, closing the discharge valve. The gas trapped in the cylinder clearance volume is never discharged but expands along the line 3 to 4.

Position 4. At this point, the pressure in the cylinder has dropped below the suction pressure, causing the suction valve to open. This opening permits a new charge of gas to enter the cylinder along the line 4 to 1.

Piston displacement is the actual volume displaced as the piston moves from position 1 to position 3. This factor is normally expressed in cubic feet per minute (cfm) and may be calculated as follows:

Single-acting cylinder (compression on one end only)

$$PD_{sa} = \frac{A_{HE} \times S \times rpm}{1728}$$

where

A_{HE} = area head end of piston, sq in.
S = piston stroke, in.
rpm = revolutions per minute
PD_{sa} = single-acting piston displacement, cfm

Double-acting cylinder (compression on both ends)

$$PD_{da} = \frac{A_{HE} \times S \times rpm}{1728} + \frac{A_{CE} \times S \times rpm}{1728}$$

$$A_{CE} = A_{HE} - A_R$$

$$PD_{da} = \frac{S \times rpm}{1728} \times (2A_H - A_R)$$

where

A_{CE} = area crank end, sq in.
A_R = area rod, sq in.
PD_{da} = double-acting piston displacement, cfm

Cylinder capacity for routine calculations can be determined by using the following equation:

$$Q = PD \times E_v \times \frac{P_s}{P_b} \times \frac{T_b}{T_s} \times \frac{1{,}440}{1{,}000}$$

where

Q = cylinder capacity, Mcf/d at standard conditions
PD = piston displacement, cfm
E_v = volumetric efficiency, expressed as a decimal
P_s = suction pressure, psia
P_b = pressure base, psia (14.7)
T_b = temperature base, degrees R (520°R)
T_s = suction temperature, degrees R
$\frac{1{,}440}{1{,}000}$ = constants to convert from cfm to Mcf/d

The definition of each of these factors has previously been given. For routine calculations, it is assumed that correction for supercompressibility is unnecessary. Where this correction is desired, it should be applied as follows:

$$Q = PD \times E_v \times \frac{P_s}{P_b} \times \frac{T_b}{T_s} \times \frac{1}{Z_s} \times \frac{1{,}440}{1{,}000}$$

Rod load is known by a variety of names, including rod load, pin load, and frame load. It is possible that this confusion of names has resulted from the fact that the weak link in a compressor cylinder design is not always the same piece of equipment. On occasion it may be the rod, or it may be the crosshead pin or bushing or the compressor frame. In any event, manufacturers of compressor cylinders will furnish the limit for it. Quite frequently the limit will be different in compression and tension, and it will usually be lower for single-acting than for double-acting cylinders. Rod loads may be calculated as follows:

$$R_c = (A_{HE} \times P_d) - (A_{CE} \times P_s)$$

$$R_t = (A_{CE} \times P_d) - (A_{HE} \times P_s)$$

where

R_c = compression rod load, lb
R_t = tension rod load, lb
A_{HE} = area head end, sq in.
A_{CE} = area crank end, sq in.
P_d = discharge pressure, psig
P_s = suction pressure, psig.

The area of the crank end can be calculated by subtracting the cross-sectional area of the rod from the area of the head end.

Discharge temperature after adiabatic compression may be calculated by using the following equation:

$$T_d = T_s R^{(N-1)/N}$$

where

T_d = discharge temperature, degrees R

T_s = suction temperature, degrees R

R = compression ratio

N = ratio of specific heats.

Utilization of horsepower

One of the most important factors in the operation of a gas compressor is the efficient utilization of gas-engine horsepower. By making maximum use of the horsepower available in a compressor unit, more gas per unit can be compressed. The result will be either handling more gas or operating with fewer units, either of which is economically desirable.

A number of factors affect compressor unit capacity and developed horsepower, such as: (1) compressor cylinder clearance, (2) suction pressure, (3) suction temperature, (4) discharge pressure, and (5) speed. Compressors are designed to utilize the engine horsepower as fully as possible for the specific conditions of suction pressure, temperature, and discharge pressure under which they are to operate. Often these conditions will change from day to day and most certainly during the life of a project. For these reasons, a certain amount of flexibility in the compressor is important.

The potential horsepower built into a compressor is fixed. The engine is designed to run at a certain speed most efficiently from the standpoints of fuel consumption and maintenance. Some speed variation is possible but usually not over a very wide range. Operating an engine at a greatly reduced speed is the equivalent of purchasing a greatly oversized piece of equipment, which is poor economy. Speeding up an engine increases its horsepower output and is one way an increase in load can be handled. It can be done safely and economically only if design load limits are not exceeded and if maintenance costs do not increase too much.

The most effective way to utilize available engine horsepower when operating conditions change is to alter the effective compressor size. The most obvious and the most expensive way to do this is to install new compressor cylinders. Other ways include the changing of cylinder displacement by boring out the cylinders or changing the liner size. One of the most common approaches is to change the compressor capacity by altering the volumetric efficiency through changes in clearance. It is common practice to provide flexibility for expected load variations by this means.

A number of methods are employed for varying the clearance in compressor cylinders. Naturally, there is a certain normal clearance built into every standard cylinder fabricated by the manufacturers. To make this cylinder fit a specific case, however, the manufacturer may alter (increase) this normal clearance to what is called a built-in clearance condition. Altering (increasing) the normal clearance is usually accomplished by one or more of the following methods.

1. Remove a small portion of the end(s) of the compressor piston.
2. Add a spacer ring between the cylinder head and the cylinder body.
3. Shorten the projection of the cylinder head into the cylinder. If the head is water cooled, the manufacturer must be consulted to determine whether sufficient metal is available.
4. Raise the suction and discharge valves by installing spacer rings under the valve.

After these changes have been made, to all intents and purposes the compressor cylinder has a new and larger inherent clearance.

Besides this basic change, there are other methods of varying cylinder clearance that are

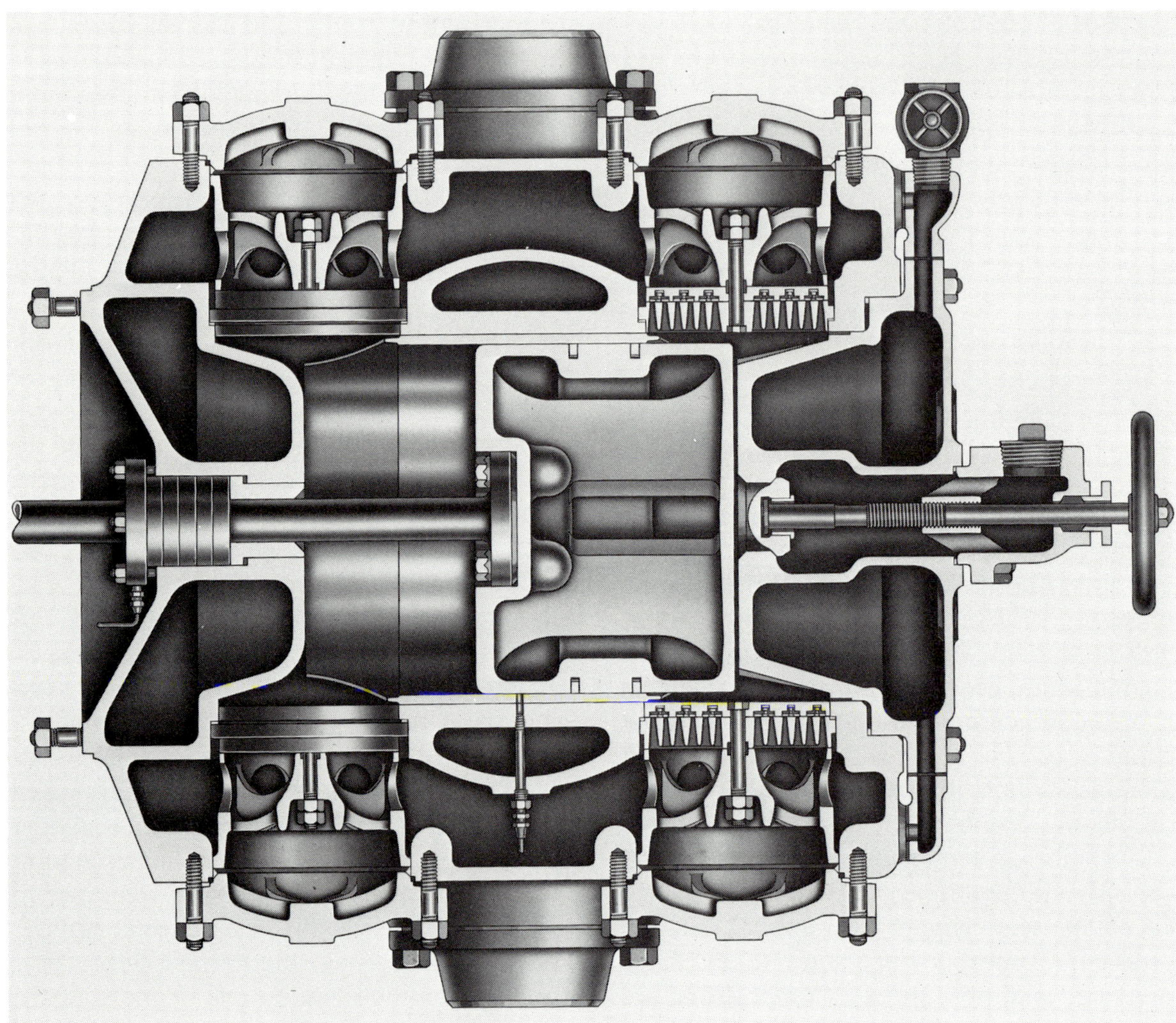

Figure 7.11. Head-end fixed-volume clearance pocket

used very frequently. Different manufacturers utilize different procedures, but the various ways may be listed and described briefly as follows:

Installation of head-end fixed-volume clearance pockets or "unloaders" (fig. 7.11) is a method that is by far the most popular and practical. Essentially, this means of adding clearance is nothing more than a fixed-volume chamber and a valve that can be opened or closed by a handwheel from outside the cylinder. Depending upon the clearance requirements as well as physical size, there may be more than one head-end fixed-volume clearance pocket built into the head-end head. Clearance pockets with variable volume are also available (fig. 7.12), but these are used less frequently because they are more expensive, maintenance costs are higher, and there appears to be little comparative advantage in their use.

Side-passage plugs are favored by some manufacturers (fig. 7.13). They are essentially flanges with plugs attached that can be inserted into passages built into the sides of a cylinder. The number of such passages as well as the size, length, and location is dependent upon the cylinder design. For some cylinders, these side passages may be located in the head-end head of the cylinder. To provide additional

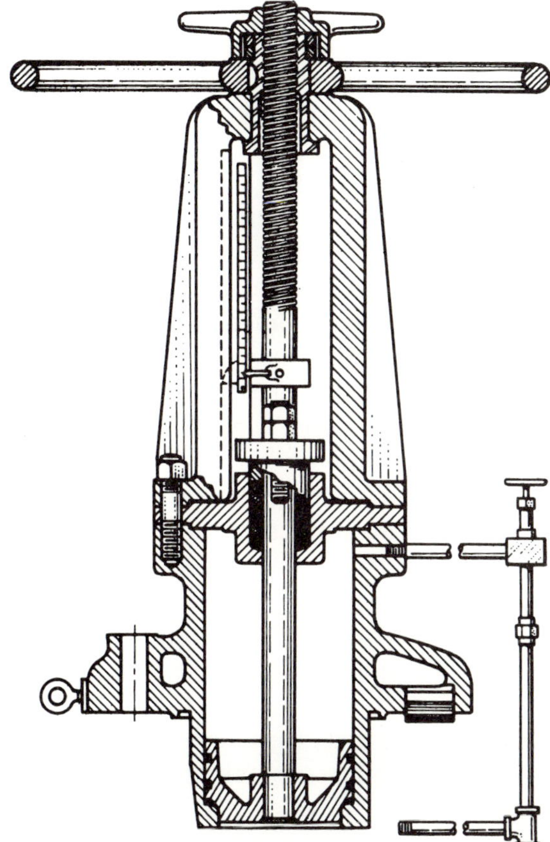

Figure 7.12. Clearance pocket with variable volume

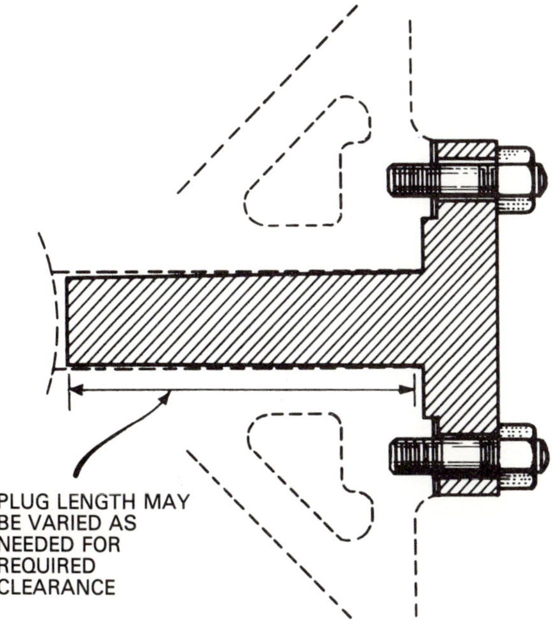

Figure 7.13. Side-passage clearance plug

clearance, the plug, or some part of it, as required, is removed from the passage.

External clearance bottles. A connection is usually provided in head-end clearance pockets to attach an external clearance bottle that provides additional clearance volume when the handwheel is opened over and above that in the pocket. For the side passages, a flanged bottle is substituted for the flange and plug discussed above when clearance volume in excess of that included in the side passages is required. There is a practical limit to how much additional clearance can be added in this manner because of the inability of the gas to squeeze in and out through the connecting passage at each stroke. One manufacturer has developed an empirical rule that limits the passage velocity to approximately 2,500 ft/min.

Crank-end (or head-end) valve cap clearance pockets. Valve cap clearance pockets may be used when it is desirable to add clearance to the crank end or when additional clearance over and above that built into the head-end fixed pocket(s) is required. A valve cap clearance pocket usually consists of a cast spherical chamber, flanged to fit in place of the usual valve cap, which can be opened or closed by means of a handwheel in the same way that the head-end fixed pocket is operated.

Suction valve unloaders (fig. 7.14) affect cylinder capacity by deactivating or lifting a suction valve that prevents that particular end of the cylinder from compressing gas. During the compression stroke, the suction valve is not allowed to act as a check valve but allows the gas to be pushed back out of the cylinder into the suction line. For large cylinders, the cylinder capacity and developed horsepower are reduced by approximately one-half.

The amount and type of cylinder clearance is dependent upon the manufacturer and the specific requirements of a job. In general, it is desirable to specify liner-type cylinders to provide flexibility in future operations and to allow for future capacity increase in absorbing the horsepower that may be obtained

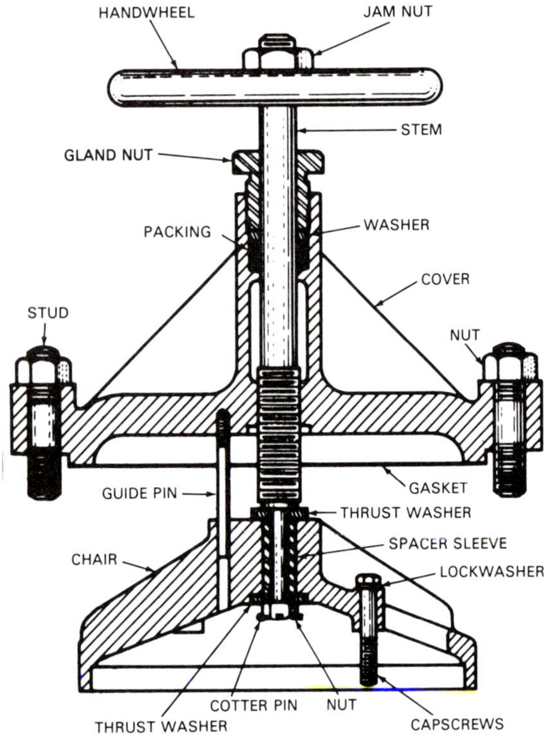

Figure 7.14. Suction valve unloader

through changing operating pressures or turbocharging an existing engine. In order to provide for flexibility in initial operations, cylinder-liner sizes and clearance arrangements are purchased in such a manner that the cylinder capacity and developed horsepower can be varied either up or down from the design point. The wisdom of providing flexibility in the manner described has been proved on many occasions, because it is difficult to accurately predict design volumes and pressures prior to plant start-up.

Compressor selection

A variety of factors influence the selection of one compressor unit over another. Among these factors are the following.

Comparative costs. Initial purchase price is certainly one of the more important considerations, and with other factors being equal, usually the most competitive offering in this respect is chosen.

Specific horsepower required versus size available. For some conditions, specific horsepower requirements are such that one manufacturer may make a unit sized very nearly to fit the need and may therefore have a competitive advantage.

Match of existing units. Matching a new unit to other equipment already existing in an established plant has many advantages, in that operating and maintenance personnel are already familiar with the operation of the proposed new unit, and parts interchangeability between units is good, making the spare part lists not as extensive.

Unbalanced loads. Manufacturers should supply information concerning unbalanced forces on the gas-engine driver; these include primary and secondary forces and both horizontal and vertical couplings. A large unbalance will require more foundation even in the best soil areas and will completely eliminate such units for marshy or water sites where the compressor unit may have to be installed on an elevated platform or a pile-supported slab.

Compressor cylinder selection and arrangement. In certain instances, manufacturers will have available units of the same general size and horsepower arrangement, but they will be different in regard to the number of compressor cylinder spots available, or they will be different in cylinder arrangement or amount of horsepower that can be absorbed in one cylinder. Although these situations do not affect unit selection very frequently, they have on occasion been a factor among others in the selection of a specific unit.

Portability. The relative portability between units is sometimes considered. For example, some types of units require two skids and others only one. Naturally, the most compact and portable unit, as long as it can still be maintained easily, will be looked upon more favorably.

Maintenance. Previous experience with different types of compressors may indicate that one type requires more maintenance attention than the other (e.g., low-speed versus high-speed units). Such evidence may have a bearing on the selection of equipment for certain services.

Life of project. The length of time that a compressor may be used at a specific location has a bearing on whether portable or permanently installed units are selected. While either type might be justified in long-lived projects, the portable type is best suited for short-term operations.

Safety considerations

In accordance with good engineering practice and as insurance against corrosion and other intangibles, all piping within the station boundaries should be designed in accordance with the ANSI *Code for Pressure Piping,* B31.3: "Petroleum Refinery Piping" or B31.8: "Gas Transmission and Distribution Piping Systems."

Special consideration should be given to the manner in which vent lines are installed so that excess pressure through improper manifolding cannot be applied inadvertently on low-pressure equipment such as starters. Also vents should be installed so that vented gas is carried well away from the unit.

Specific detailed starting and stopping instructions should be posted adjacent to the units, and operators should be required to follow such instructions.

In the design of any installation to be operated largely unattended, it is necessary to incorporate certain features of automatic control for protecting equipment in case of malfunction. For protection, all engines are equipped with standard shutdown devices for low lube-oil pressure, overspeed, and high jacket-water temperature. The engine can be shut down either by grounding the magneto or by blocking and venting the fuel. The latter arrangement has the advantage that fuel gas cannot accumulate in the exhaust and muffler where it might subsequently detonate or cause a backfire. Additional features may be incorporated at the discretion of the designer to conform with company standard practice or local safety regulations.

A scrubber should be installed immediately upstream of each stage of compression and should include an automatic drain and a device for shutting down the compressor in the event of a high liquid level in the scrubber.

Since a compressor cylinder is designed to handle a certain volume of gas under specific conditions of speed and suction pressure, it is essential to control the pressure within certain limits. For example, if the suction pressure is too high, overloading the compressor frame or driver is possible as a result of the compressor cylinder handling more than the design volume of gas; if the suction pressure is too low, the discharge temperature of the gas from the compressor may rise above safe limits because of the heat of compression, resulting in valve packing, rod, or piston failure. Consequently, it may be desirable to install a high-discharge temperature shutdown, particularly if the compressor cylinder is equipped with only one discharge valve on each end. For additional safety, a high-discharge and low-suction pressure shutdown may be provided to function in the event of a gathering line rupture or a station discharge line pressure increase above design limits.

One common practice in connection with control of suction pressure is to vary the speed of the driver automatically to maintain a reasonably constant pressure. With or without speed control, a cycling regulator may be used to divert gas from the discharge side of a compressor and throttle it back into the suction side, thus maintaining a minimum suction pressure to smooth out the engine operation and to assure gas circulation. This procedure has the disadvantage of wasting horsepower, but in many instances it is required to maintain satisfactory compressor operation.

Another practice for preventing excessively high suction pressure is to employ either a pressure-reducing inlet regulator or a back-pressure regulator to flare. In most casinghead gas gathering systems, either method is satisfactory, but in compressing gas-well gas, the inlet regulator should be used in cases where the gas may not be flared. On installations of the latter type, the wells should shut in automatically, or the gathering system should be designed for the maximum wellhead pressure.

As a supplement to the usual automatic controls, it is considered wise to have one or more manually operated shutdown stations strategically located safely away from the station. In case of an emergency, these controls can be used to shut down the compressor from a safe distance.

Turbine-driven centrifugal compressors

Gas turbine engines

A gas turbine engine provides a constant flow of power, contrasted to pulses supplied by a reciprocating engine, by passing hot gases at high velocity across the blades of a turbine wheel. The shaft to which the turbine wheel is attached delivers the power in the form of rotational energy. The basic components of an internal-combustion engine are shown in figures 7.15, 7.16, and 7.17 and are listed below.

1. The compressor. The axial or centrifugal flow compressor provides air under pressure for combustion in the combustor section and, as necessary, for cooling the engine.
2. The combustor. Air from the compressor and fuel at a pressure of about 150 to 200 psi from a fuel pump or gas supply line are brought together and mixed in this part of the engine, and combustion takes place. The hot gases flow to the gas-producer turbine wheel, which is attached to the same shaft as the compressor.
3. The gas-producer turbine wheel. The hot gases passing through the blades of the gas-producer turbine wheel cause it to rotate, driving the compressor and other accessories such as the fuel pump and oil pump.
4. The power turbine. The hot gases then pass over another set of turbine blades attached to a shaft. This may be an extension of the shaft to which the gas-producer turbine is attached, or it may be entirely independent. This is the power turbine that provides the power to operate the machine that the engine is designed to drive.
5. Exhaust gases. The hot gases then pass out the exhaust. The remaining heat energy is often used in boilers, oil heaters, and so forth to recover heat that would otherwise be wasted.

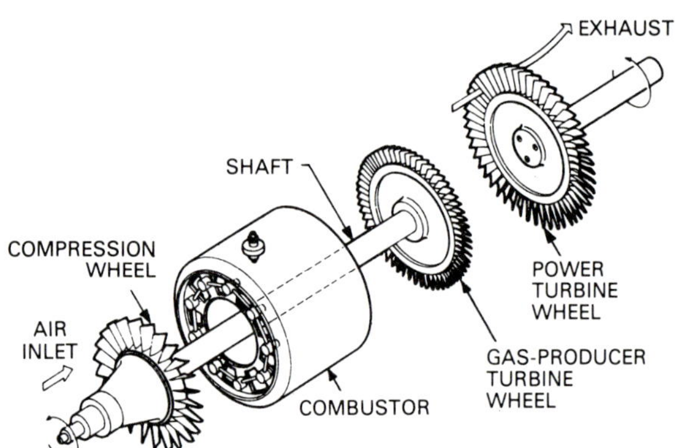

Figure 7.15. Main components of a gas turbine engine

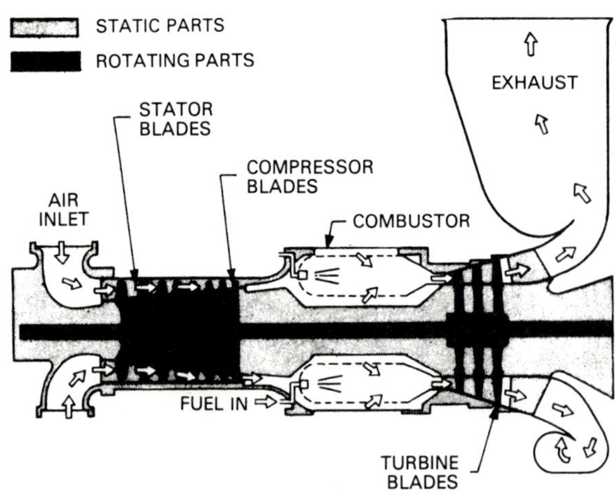

Figure 7.16. Air and gas flow in a gas turbine engine

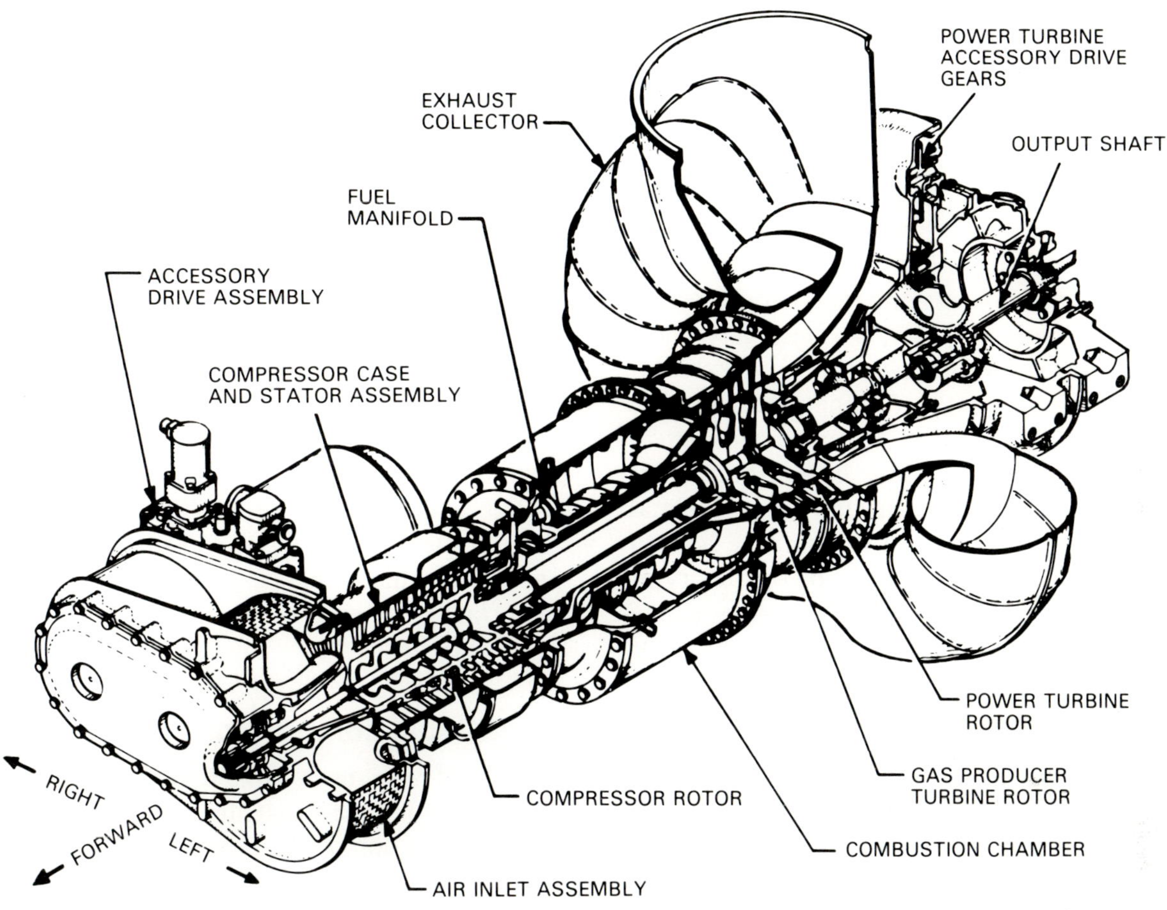

Figure 7.17. Cutaway drawing of a gas turbine engine

Large amounts of air are required by turbine engines in order to control the temperatures in the combustor and turbine sections. About four times as much air is used as would be required for stoichiometric combustion—complete combustion with no excess oxygen. By this means, firing temperatures are held to a range of 1,450°F to 2,200°F. Most industrial turbines operate in the lower part of this range, while aircraft turbines operate in higher temperatures. Exhaust-gas temperatures may run as low as 850°F.

Gas turbine engines are designed for a wide range of speeds—14,000 to 23,000 rpm being common. They can also operate efficiently at other than design speeds, within reasonable limits. Gas turbine engines can be directly connected to a high-speed compressor, or the power output can be reduced to a lower speed through gearing for driving a pump or for other low-speed applications.

Gas turbine engines can operate on kerosine or diesel oil, natural gas, butane or propane, or other fuels. Dual systems are sometimes used for distillate and natural gas. These units adjust easily to changes in the heating value of fuels, since the power output is a direct function of the Btu's released in the combustor. Thus, a reduction in heat value of the fuel will simply result in the fuel throttle opening up to admit a greater amount of fuel. Fuel consumption usually approximates 10,000 – 13,000 Btu/hp/h for optimum operating conditions.

In a reciprocating-type engine, power is obtained by burning fuel, and the resultant increase in the volume of gases causes an increase in pressure in the power cylinder. This pressure, or force, pushes the piston, and force (F) moves through a distance (D) and results in work (W). Remember, $W = F \times D$. In the case of the turbine engine, fuel is burned in the combustor, resulting in a large increase in the volume of gases; as these gases flow through the turbine wheels to the exhaust, they impinge on the turbine wheel blades, exerting a force against them and causing the wheels to rotate like fans in a strong wind. Again, a force moves through a distance and results in work. It is not the purpose here to provide a detailed discussion of the theory of gas turbine engines; such information can be obtained from manufacturers of these units or from standard engineering texts.

The starting sequence for gas turbine engines is automatic once the operator actuates the starter button. The engine is rotated by an outside power source such as an electric, hydraulic, air, or gas expansion motor for a short period of time to purge the turbine and exhaust system and to place all lubricating systems in operation. Rotation at this point is relatively slow. Next, fuel is supplied to the combustor and is ignited. The unit then comes up to speed, and the load is applied. Only a minute or so is required for this entire operation.

Elaborate controls and protective devices are provided to protect against high exhaust-gas temperature, low lube pressure, high lube temperature, low lube-oil level, and turbine overspeed. Control systems are available to permit the completely automatic operation of unattended units in remote locations.

Maintenance requirements on the main part of the gas turbine engine are relatively minor if the air and fuel entering the unit are free of dirt and water. It can be seen readily that such impurities would damage the turbine blades and otherwise adversely affect the close non-wear tolerances necessary for efficient operation. Adequate filtering is absolutely essential. With filtering, major reconditioning of the turbine internal parts needs to be carried out only after intervals of 25,000 to 40,000 hours of operation. Maintenance on the auxiliary equipment and controls needs to be performed at much more frequent intervals, particularly for remote automatically operated units.

Although shaft speeds are high, bearing loads are low due to the steady power flow and lack of load reversals. For these reasons, lubricating problems are not great and oils generally have a long service life. However, care should be taken to keep the oil clean and within specification limits. To accomplish this, frequent testing of the lube oil should be part of the maintenance program. The unit manufacturer will provide the lube-oil specifications.

Centrifugal compressors

One of the laws of physics (Newton's second law) states that force equals mass times acceleration. Force is ordinarily expressed in terms of pounds. However, pressure, or pounds per square inch, is also a measure of force. This principle is used in centrifugal compressors. Power from a prime mover is used to rotate impellers, which contain blades. The rotating impellers accelerate the gas to a given velocity before the gas leaves the impellers. The stationary diffusers and guide vanes convert the gas velocity leaving the impellers into pressure.

A centrifugal compressor consists of an outer steel housing containing the internal stationary parts—diaphragms and guide vanes—and a series of impellers mounted on a shaft, located within the housing with the impellers properly aligned with respect to the diaphragms (fig. 7.18).

Centrifugal compressors generate a relatively low pressure ratio per stage of compression. This is a desirable characteristic for many natural gas transmission applications where a pressure ratio of 1.25 is typical to boost gas

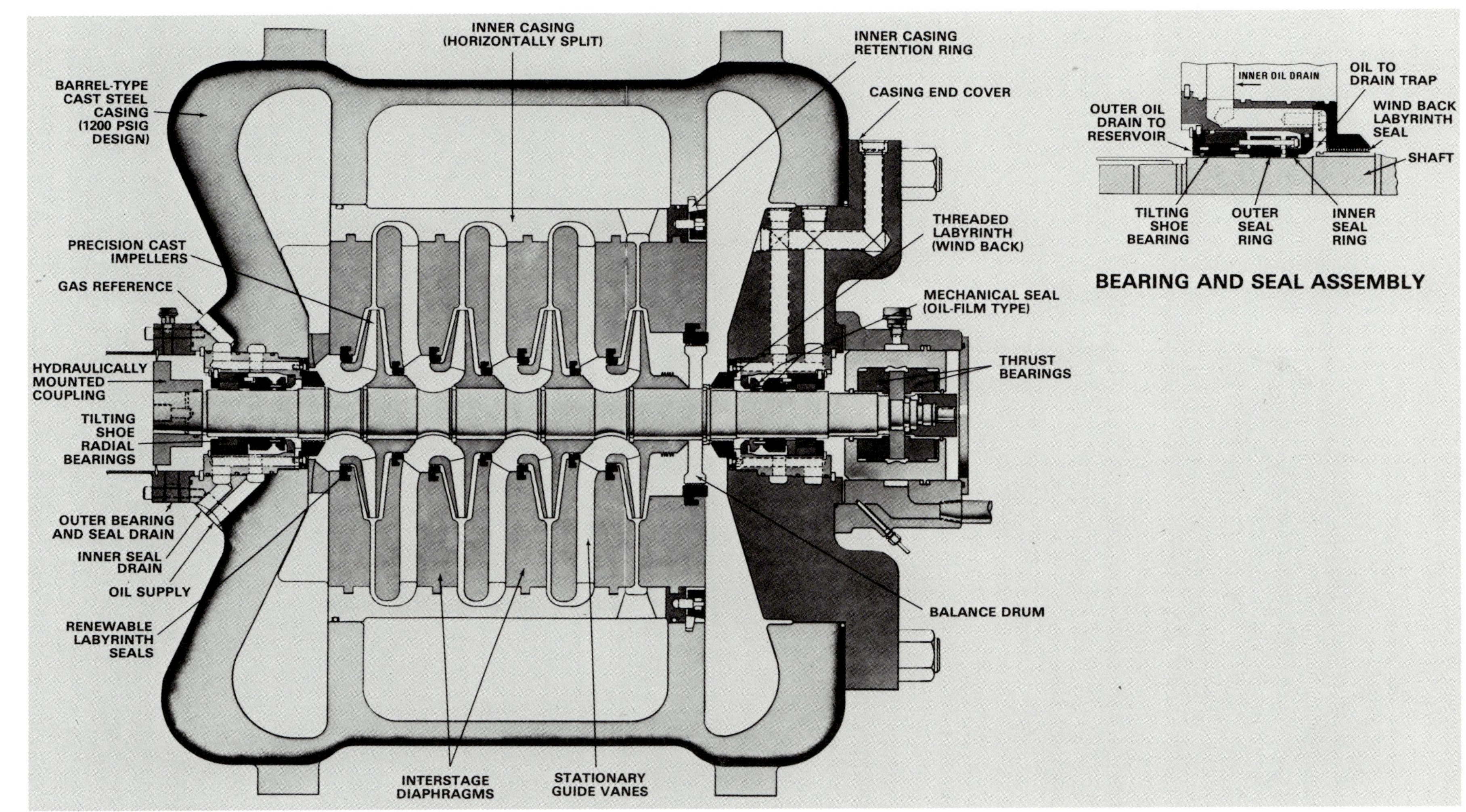

Figure 7.18. Cross section of a centrifugal compressor

from 800 psi pressure to 1,000 psi. When higher pressure ratios are required, stages are run in series, as shown in the four-stage compressor cross section (fig. 7.18). Gas transmission compressors vary in power and speed from 1,000 hp and 20,000 rpm to 30,000 hp and 3,600 rpm.

Natural gas gathering applications require much higher pressure ratios than gas transmission. Pressure ratios of 30 are typical to compress gas from 50-psi pressure to 1,500 psi. To achieve this kind of ratio, several casings containing up to eight stages each are run in series. Gas reinjection compressors have been built for pressures up to 10,000 psi.

Figure 7.19 illustrates the performance of a typical multistage centrifugal compressor. It should be noted that, along a constant speed line, as the flow is decreased, the head or pressure ratio increases until a point is reached where stable operation is no longer possible. This is commonly called the *surge limit* and is

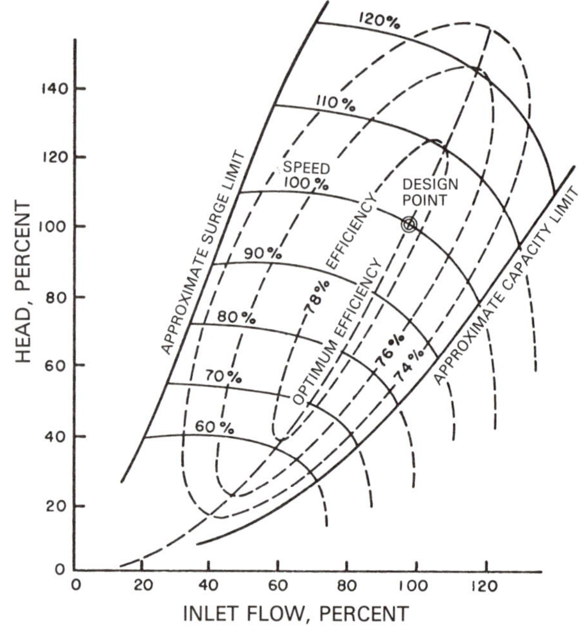

Figure 7.19. Multistage centrifugal compressor performance curves

Figure 7.20. 5000-hp gas turbine driven centrifugal compressor

characteristic of all centrifugal compressors. Operation in this region must be avoided and is usually prevented by a surge control system.

The advantages of centrifugal compressors are their low maintenance and minimum installation requirements. Their simple construction, with the rotor being the only moving part, results in a very reliable machine. There are no valves or pistons to produce unsteady reversing loads. Foundation and support requirements are minimal. Continuous-operation runs of five years and longer are not uncommon.

Figure 7.20 shows the installation of a 5,000-hp gas turbine driven centrifugal compressor in gas transmission service.

8 Instruments and controls

Important variable quantities are associated with the production and handling of natural gas and any associated liquids—pressure, temperature, flow rate, and liquid level. How these variables are controlled is crucial to safety, operability, and accurate measurement of quantities of natural gas and associated liquids.

Some of the terms associated with instruments and controls should be learned. The *controlled variable* in a system is the variable whose value is the objective, the aim, of the control system. Its correct value is called the *set point* value. In field handling of natural gas the controlled variable may be any of the four mentioned in the paragraph above. Sometimes one or more of these variables may be used to aid in the control of the controlled variable. Used in that way they become *manipulated variables.* Should steam be used to heat water in a hot water system, *temperature* of the water is the *controlled variable,* steam is the *control agent,* and *flow* of steam the *manipulated variable.* Additionally, the valve that controls flow of steam is a *final control element,* the device that initially measures the water temperature is the *primary element,* and the hot water is the *controlled medium.*

Modes of control

Mode of control refers to the manner in which a manipulated variable is handled in order to regulate the controlled variable. By far the most commonly used mode is *on-off* action, in which the flow of the manipulated variable is at a maximum or at zero. The controls on domestic water heaters, heating and cooling systems are usually of this sort.

The *proportional* mode is popular in many process control systems. It provides continuous flow and responds to changes in the controlled variable, but it is not necessarily capable of maintaining the controlled variable at the set point.

Reset and *rate* modes usually work in conjunction with the proportional mode. Reset mode working with proportional mode (proportional-plus-reset) is capable of rather quickly returning the variable to the set point. Rate mode is used in process control in conjunction with proportional and reset (proportional-plus-reset-plus-rate). This arrangement provides a finer control, allowing less deviation from the set point, and generally a more rapid return to it. It is not needed in the vast majority of cases.

Measurement of controlled variables

Before a variable can be controlled its value must be measured or compared with its desired value in order to obtain an *error signal.* The error signal is the difference between the desired and the actual values of the controlled variable and is used to initiate corrective action by the controller. A large number of devices and instruments are capable of measuring or comparing the values of the several variables that are of interest in field handling of natural gas.

Pressure measurement

The *manometer* is a common instrument for measuring rather low values of pressure, from those well below zero gauge pressure to perhaps a few pounds per square inch gauge (psig). The manometer is typically a U-shaped device made of glass tubing (fig. 8.1). The

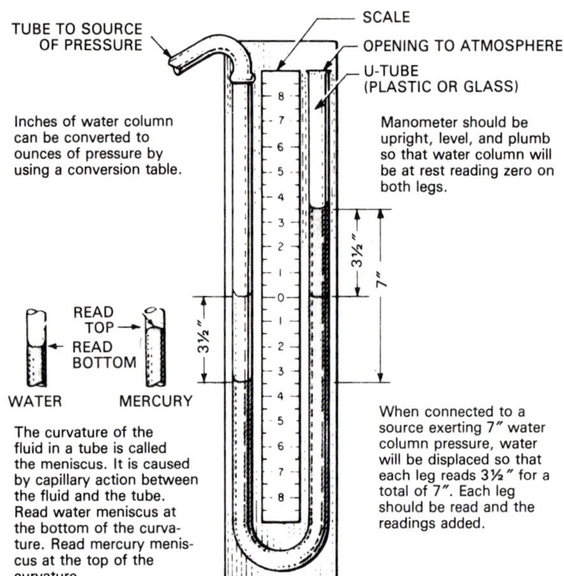

Figure 8.1. Typical water manometer

inside diameter of the tube is unimportant, but it must be consistent over the entire tube length. A scale with suitable markings is attached behind the unit and the tube filled approximately half full with an appropriate liquid. The scale will be marked for a liquid having a particular specific gravity. One end of the tube is open to the atmosphere, and the other end is connected to the source of pressure to be measured. The pressure will force the fluid to move in the tube until the weight of the displaced fluid is equal to the force exerted by the pressure. Difference in height of the liquid in the two sides of the manometer is an indication of the pressure. Pressure measured by a manometer is usually expressed in millimetres of mercury, inches of mercury, or inches of water column. Manometers find little use in field handling of natural gas.

A *diaphragm element* is also used in measuring low pressure values, up to a few pounds per square inch gauge (fig. 8.2). Diaphragm units are found in some control devices and in special instruments such as barometers which are commonly used to monitor atmospheric pressure. Sometimes diaphragm units are stacked, with three or more connected in series (fig. 8.3), affording greater movement and thus bringing about greater sensitivity.

A *bellows unit* is similar to a diaphragm element. The bellows is usually manufactured from a thin seamless tube of selected material. The tubing is subjected to a machine that exerts pressure to stretch and form the metal into the convolutions, the name given to the corrugated effect seen in figure 8.4. Base plates and other fittings are soldered, brazed, or welded to the unit. A bellows unit may

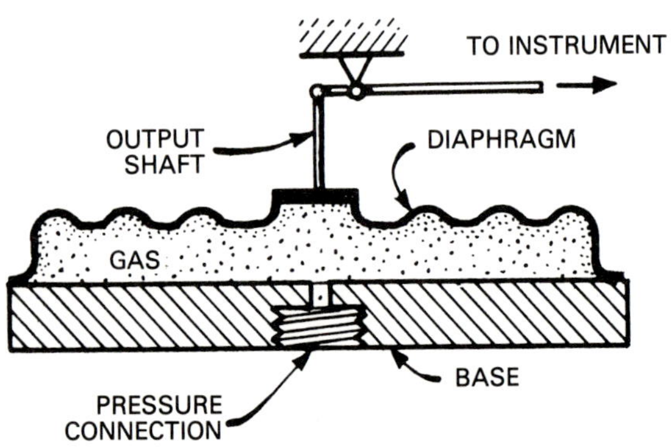

Figure 8.2. Single-diaphragm pressure-sensing element

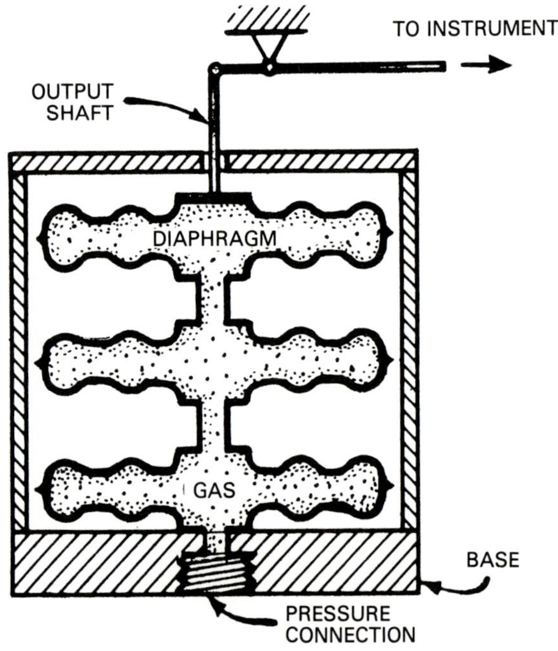

Figure 8.3. Multiple-diaphragm pressure-sensing element

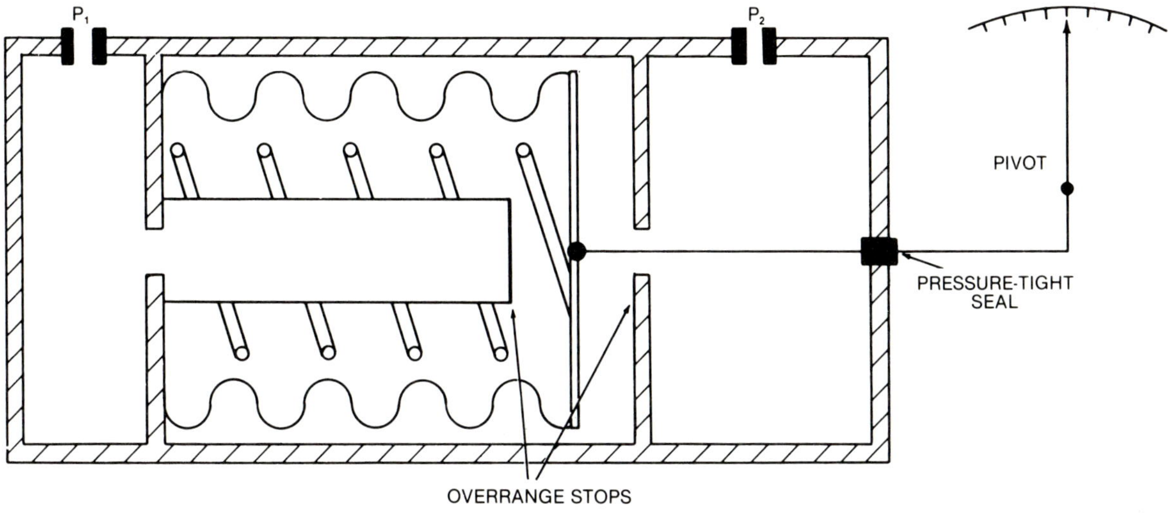

Figure 8.4. Bellows with a load spring, used to measure pressure

have pressure applied internally or to the outside surface. Pressure applied at p_1 in the figure will cause the bellows to expand or lengthen; applied to p_2, it will compress or shorten the unit. For measurement and control purposes bellows units are available in a variety of sizes and are generally capable of handling pressures of 30 psig or more.

A *Bourdon tube* is commonly a C-shaped metal tube sealed at one end, called the free end or tip, and a pressure connection at the other end (fig. 8.5). Pressure applied at the connection will exert a force on the entire inner surface of the tube. Since the outer circumference is greater than the inner, this force will tend to straighten the tube, causing movement of the free end or tip. This movement is used to position an *indicator needle,* a *recording pen,* or the *flapper* in a flapper-nozzle arrangement of a controller. The amount of movement of the tip end is proportional to the applied pressure.

The C-shaped Bourdon tube may need a sector gear and pinion to amplify its movement. The pinion will drive a pointer that can be arranged to cover a circular scale of more than 300°. Another form of Bourdon tube is one wound to form a helix (fig. 8.6). This version will give far greater movement for a given

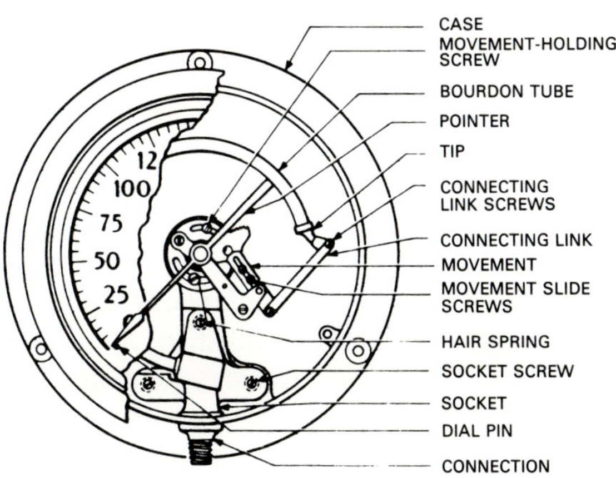

Figure 8.5. Bourdon tube pressure-indicating instrument

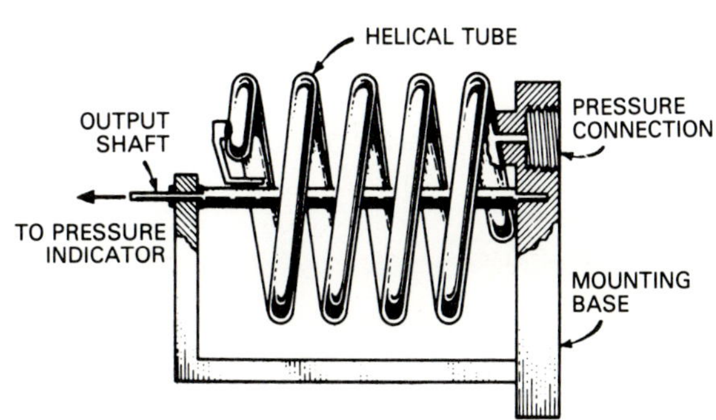

Figure 8.6. Helical tube pressure-sensing element

pressure application than the C-shaped version. Such elements avoid some of the frictional and other possible errors that can exist in gauges driven by C-shaped elements. Both C-shaped and spiral Bourdon devices can be designed to handle pressures in the thousands of pounds per square inch range.

Temperature measurement

A filled glass tube thermometer is the most widely used instrument for measuring temperature. A glass tube with a very small interior column, sealed at the top and having a bulb at its bottom end, makes up the major part of a common thermometer. Its bulb will be filled with a liquid during manufacture. The liquid may be mercury, pentane, toluene, or alcohol, depending on the range of temperature to be measured. All these liquids expand and contract according to their temperature. If a scale is now etched or otherwise placed on the tube above the bulb, the thermometer is essentially complete.

Thermometers or other temperature-sensitive elements are usually placed in a thermometer well when it is necessary to measure temperature in pressurized vessels or pipes (fig. 8.7).

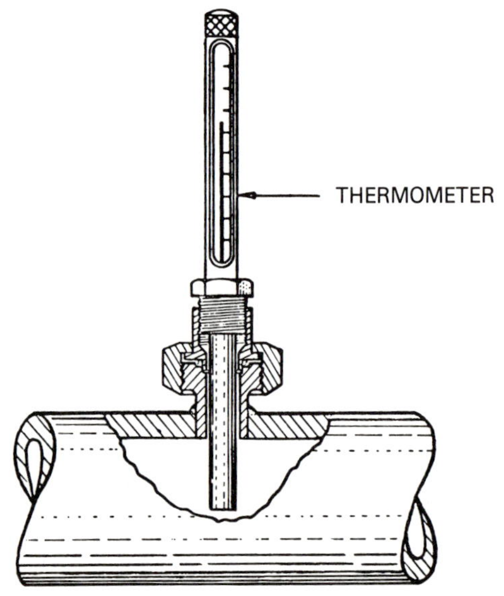

Figure 8.7. Thermometer well

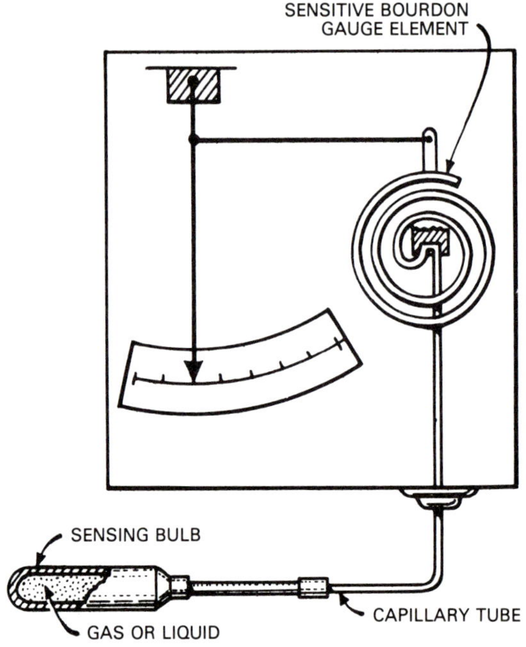

Figure 8.8. Filled-system thermometer

A *filled system* thermometer (fig. 8.8) provides mechanical force to position an indicating pointer, recording pen, or error-detecting unit of a controller. The system consists of a gas- or liquid-filled bulb, small-diameter tubing, and spiral Bourdon tube element. Such systems depend on expansion and contraction of the fluid in the bulb to provide driving force to the Bourdon tube.

The ambient temperature will affect the fill medium in the connecting capillary tubing and the spiral Bourdon tube element. To achieve accurate response to the bulb temperature it is necessary to compensate for varying ambient temperature. Excellent compensation can be had by incorporating compensating tubing and a compensating spiral element (fig. 8.9). The compensating Bourdon tube element is spiraled in a direction opposite to the measuring element. Any movement of the measuring element due to a change in ambient temperature will be exactly offset by action of the compensating Bourdon tube.

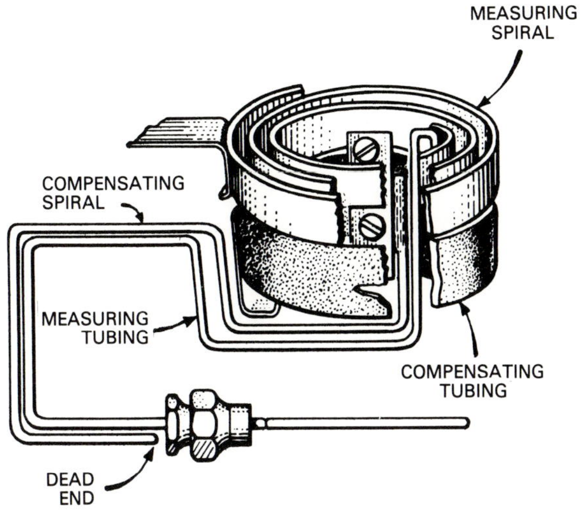

Figure 8.9. Compensating filled-system thermometer element

Some filled systems contain a partial filling of a fluid and depend on the change in vapor pressure caused by a temperature change. This change in vapor pressure is a good indication of temperature change at the filled bulb, and the method has the advantage of not being affected by ambient temperature changes. A disadvantage lies in the fact that measuring range is limited for each of the common fills used (ethyl alcohol, ether, propane, butane, and others). For proper results the boiling point of the fill fluid must be lower than the lowest temperature to be measured.

A *bimetal temperature element* is made up of two strips of different metals. All metals will expand or contract with temperature changes, but the rate of expansion from one metal to another will differ. If two dissimilar metal strips are bonded together along their length and held rigid at one end, the free end of the unit will move as the temperature changes. At some given temperature the unit will be straight, but an increase in temperature will cause the free end to curve in one direction, and a decrease in temperature will cause it to curve in the opposite direction (fig. 8.10).

Free-end movement of a bimetal element is directly proportional to its length. To conserve space most bimetal elements are wound

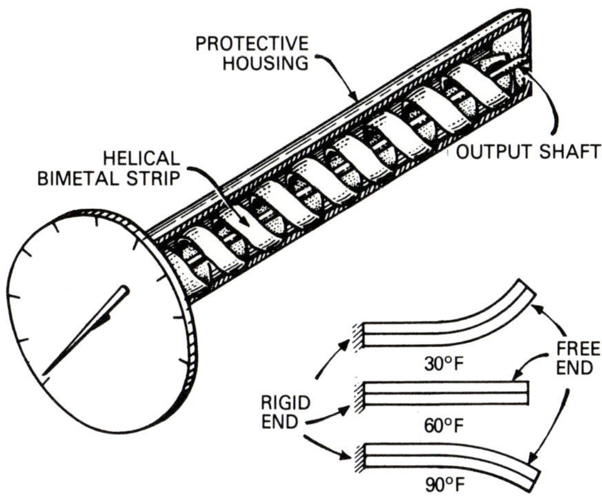

Figure 8.10. Bimetal element for temperature measurements

in a spiral or helical shape. Bimetal elements find extensive use in instrumentation.

A *thermocouple element* consists of two wires of different metal bonded to form a junction (fig. 8.11). Heating this junction will produce a potential (a voltage) that can be measured between the free ends of the two wires. It is necessary that the junction have a

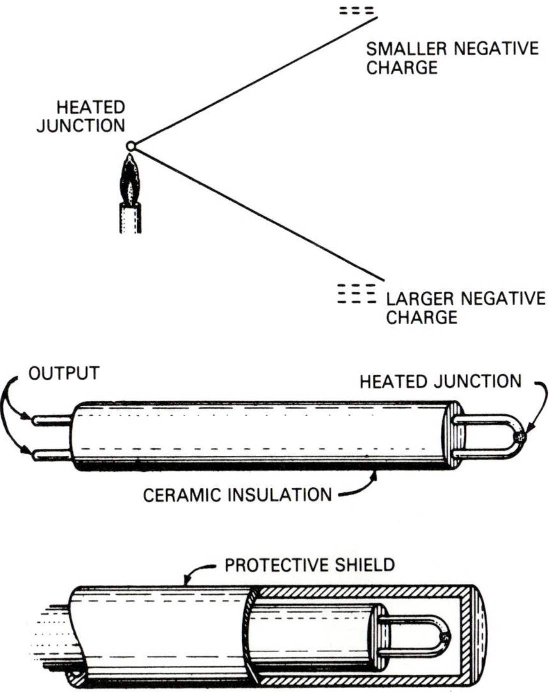

Figure 8.11. Thermocouple temperature element

temperature different from that existing at the other ends of the two wires in order to have a measurable potential. At best the potential is feeble, but a millivoltmeter can measure it, and its value will represent an accurate account of the temperature at the junction. More precisely, the potential will represent the difference in temperature between the thermocouple junction and the junction at the other ends of the wires.

Thermocouples are used to measure various ranges of temperature, the upper limits being those that would tend to destroy the elements. Thermocouples are not used extensively in oil and gas field instrumentation. One use that can be found, however, is as a safety shutdown feature on some gas-fired equipment. The thermocouple junction is heated by a gas pilot light, and output from the thermocouple energizes a solenoid switch that can interrupt current flow to a solenoid valve located in the gas line to the burner. Should the pilot light be extinguished for any cause, current from the thermocouple is lost, so the solenoid valve closes, preventing the escape of gas or a worse situation.

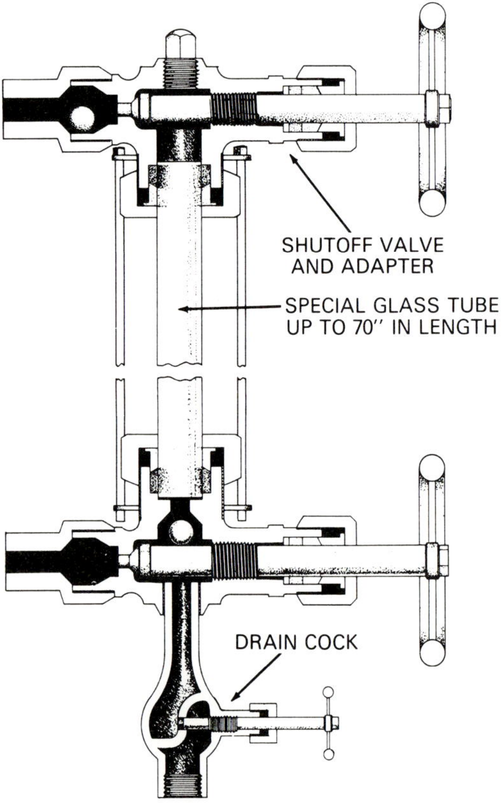

Figure 8.12. Sight glass for observing liquid level in a vessel

Liquid-level measurement

A common device for direct reading of liquid level is the *sight glass.* It consists of a glass tube, upper and lower shut-off valves, and adapters. Such liquid-level indicators can be adapted to fit the vessel whose liquid is to be monitored. For use on pressurized vessels the device may contain metal balls and seats that will act to shut off leakage should the glass column break (fig. 8.12). Tubular glass for this sort of gauge is available in lengths to 6 feet and pressures to 600 psi. Armored sight glasses are available for much higher pressures.

A *float-type unit* is sometimes used when liquid-level control is needed. One such unit uses a hollow sphere that floats on the surface of the liquid (fig. 8.13). In *A* of the figure, the float drives an indicator device for the full desired range of liquid level. In *B* the float is connected through a lever system to a valve. Motion of the float is limited by the amount of valve stem movement.

A commercial form of liquid-level controller is designed to be mounted in a flanged hole of a pressurized vessel or mounted in a float chamber outside the vessel (fig. 8.14). Motion of the float is clearly limited to small range but is sufficient for most control purposes. Motion of the float turns a shaft that protrudes from the flanged housing through a stuffing box. Linkage between the rotary shaft and the valve can be arranged to open or close the valve with a rise in liquid level.

The force exerted by a liquid in a containing vessel is directly proportional to the depth of the liquid at the point of measurement. This force will produce a pressure, and if the pressure can be measured, the depth can be determined. A sensitive bellows or diaphragm element can be used to measure the pressure

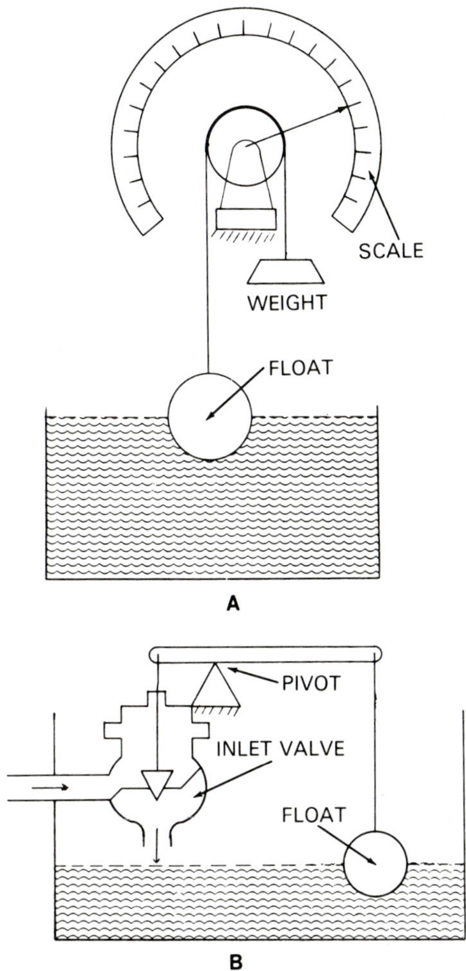

Figure 8.13. Float-type instruments with (*A*) full-range indicator and (*B*) restricted-range controller

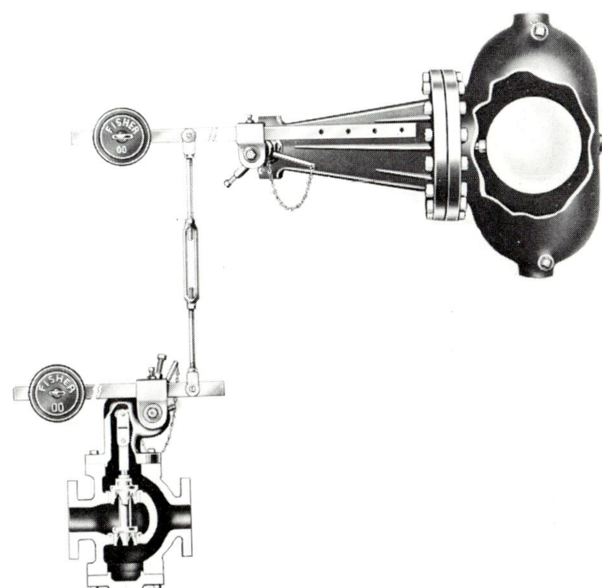

Figure 8.14. A float-operated liquid-level controller mounted on a float chamber (Courtesy of Fisher Controls)

(fig. 8.15). This device will function in either an open container or one that is pressurized.

That part of the mechanism containing the process diaphragm is flange-fitted to the vessel containing the liquid, perhaps flanged to the bottom of the vessel. If not, it will probably be flanged to the side of the vessel very near the bottom. Should the device be used on an open tank or vented vessel, the opening for the compensating connection will be vented to the atmosphere, usually through a filter.

For open tank use, pressure on the process diaphragm will be directly proportional to the liquid level in the tank. This same pressure will be transmitted to the primary diaphragm through the liquid fill that connects the two diaphragms. Note that the primary diaphragm has a liquid fill on both sides. Fill on the right side connects to the compensating diaphragm, but for open tank use, this fact has no effect on operation of the unit.

The primary diaphragm drives a push rod and an output shaft. The output shaft passes through a flexure seal, which also acts as a

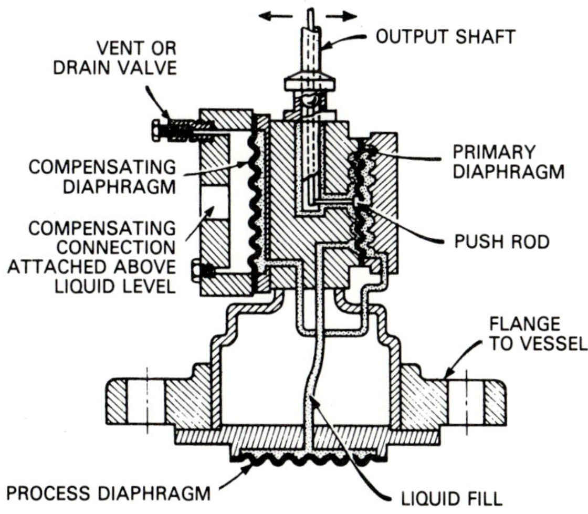

Figure 8.15. Diaphragm mechanism for a pressure-compensating liquid-level controller

pivot point. Upper end of the output shaft may be connected to a measuring means or made to operate a flapper-nozzle pneumatic control element.

When this device is used on a pressurized vessel, the process diaphragm will have two pressure sources acting on it. One source is due to the liquid level, and the other source to the added pressure that exists above the surface of the liquid. Values of the two pressures add together, and the sum of their forces acts on the process diaphragm. The sum of these forces is transmitted to the left side of the primary diaphragm.

The pressure that exists above the surface of the liquid is piped to the compensating diaphragm, where it exerts a force that is transmitted through the liquid fill to the right side of the primary diaphragm. At this point it acts in opposition to the forces on the other side of the primary diaphragm. The force due to the added pressure above the liquid is exactly canceled out, leaving only the effects of pressure due to liquid-level height.

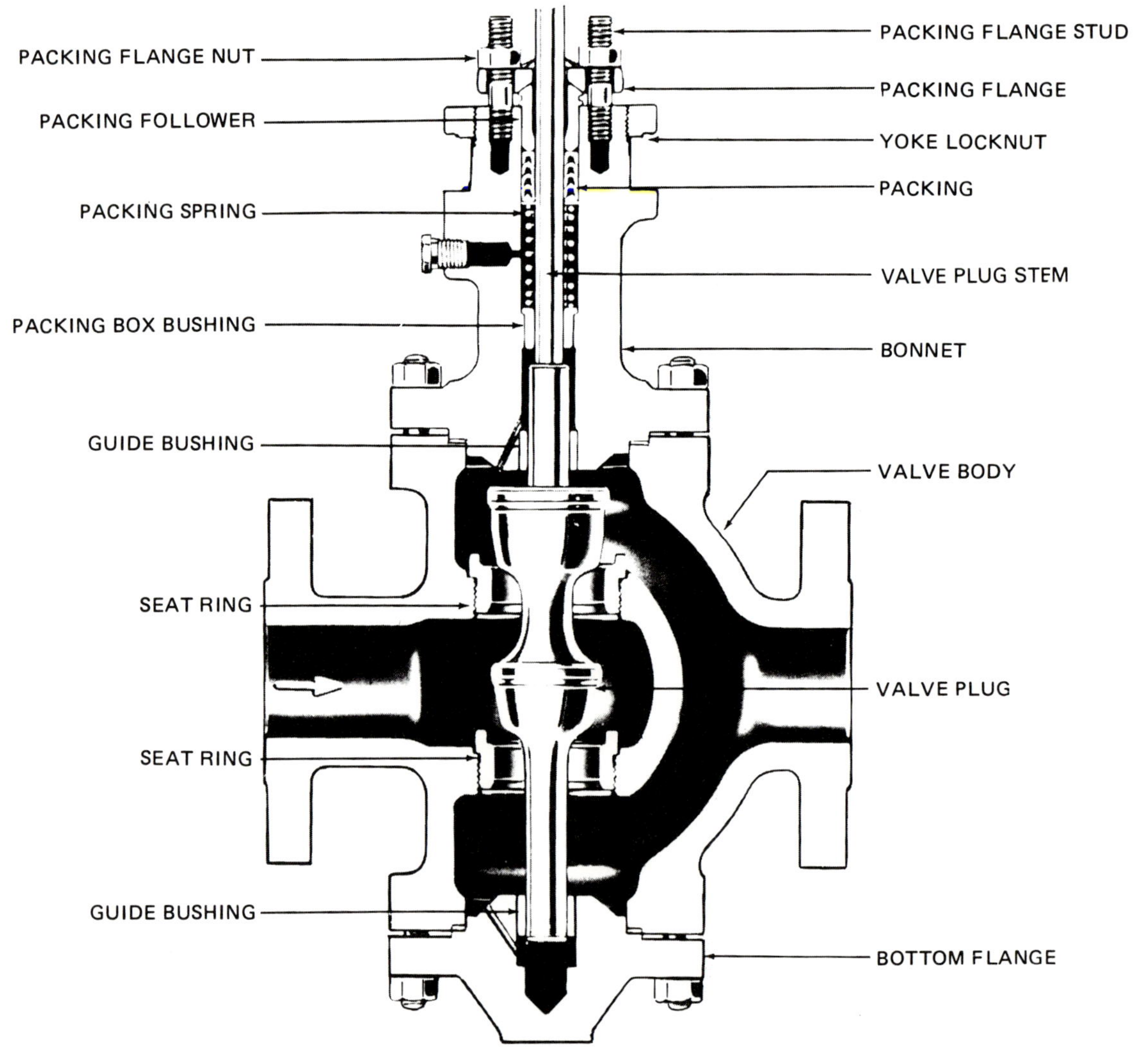

Figure 8.16. A double-ported valve, showing nomenclature commonly used for various parts (Courtesy of Fisher Controls)

Valves

Valves serve to control the flow of fluids, and generally are able to shut off flow entirely and to adjust flow to the maximum rate possible. The main components of a valve are its *body, plug, guides,* and *seat;* other components as well are shown in figure 8.16. Most valve bodies are globes, but other shapes include split-body, angle body, and a form used in three-way valves. Valves are connected in flow lines by any of several methods: (1) threaded connections are in common use for relatively small sizes and moderate pressures; (2) welding is used to attach very large valves, particularly in pipeline service, as well as small valves in some services; (3) flanged connections are common for small to large sizes and for high pressures such as might be found on a wellhead.

Single-ported valves

Single-ported valves have a single path for fluid flow (fig. 8.17); the two forms shown are similar but have important differences. Note the methods of attaching bonnets and bottom flanges. Example *A* uses studs and nuts, while *B* uses clamp rings. An important difference is that plugs are reversed with respect to one another. Pushing down on the stem of valve *A* causes it to close, while the same action on *B* causes it to open. Valve *A* is called a *direct-acting* valve, and *B* is a *reverse-acting* valve. Each form finds wide use in control applications.

Close examination of figure 8.17 reveals that plugs in these valves can be reversed, thus changing the valve from direct-acting to reverse-acting, and vice versa. The valve seats must also be reversed, of course. An advantage of this feature is reduced manufacturing cost, since a single set of parts can provide either a direct- or a reverse-acting valve.

Single-ported valves have advantages over double-ported valves. They cost less, are easier to maintain, and are less susceptible to leakage when fully closed. Line pressure can adversely affect valve plug movement when the plug is in the closed or near-closed position. Smoothest operation is obtained when the valve is installed so that line pressure tends to force the plug away from its seat. When

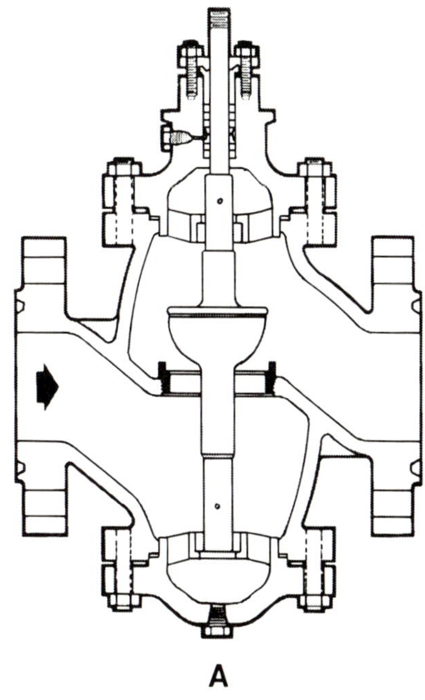

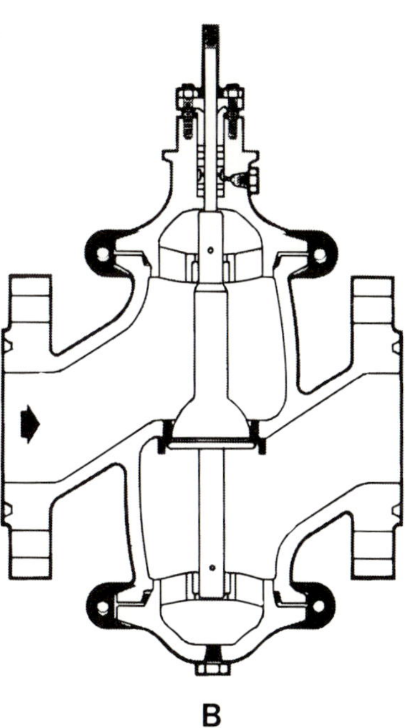

Figure 8.17. Two types of single-ported valves. *A*, direct-acting; *B*, reverse-acting.

used in throttling service, single-ported valves serve well if the actuator is aided by a valve positioner.

Double-ported valves

Double-ported valves (fig. 8.16) are popular for automatic control applications. Compared with single-ported valves, the double-ported valves require considerably less force to position the plugs. Pressure acts on the plug at both seats, tending to force it open at one and to force it closed at the other. These opposing forces tend to cancel each other to the extent that the plug moves with greater ease than that for the single-ported valve.

Gate valves

Gate valves are found on wellheads as well as in other service. Such valves (fig. 8.18) are meant to be operated at either full flow or complete shutoff. They are described as full-opening, through-conduit valves, and as such offer considerably less resistance to flow than any other form of valve. The through-conduit feature is especially important on wellheads where wireline tools may occasionally need to be inserted into the well. Most of the valves on a wellhead installation are manually operated. The safety shutdown valve is usually a spring-loaded, pneumatically actuated gate valve.

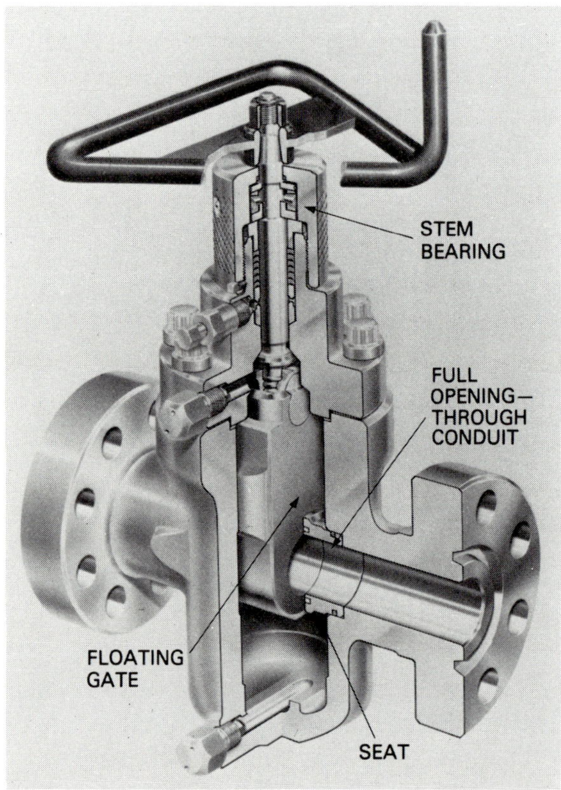

Figure 8.18. Gate valve typical of those used on wellheads

Valve characteristics

Valve characteristics must be considered when selecting a control valve for a given situation. Principal characteristics are related to shutoff capability and the flow characteristic of the valve. Some valves are designed to be opened all the way or to be shut off completely. Others are intended for throttling service and may not provide a tight shutoff. Usually the kind of valve plug used will determine the sort of service for which the valve is best suited.

Valve plugs

Flow characteristics of a valve describe the relationship between the amount of fluid flow through the valve and the extent to which the valve is open. A valve having a *linear flow characteristic* is one having a flow rate directly proportional to the amount of opening. The flow characteristic is determined largely by the shape of the plug and seat arrangement. The *quick-opening plug* (fig. 8.19) achieves a high rate of flow with little opening. When the plug is 50% open, the flow rate is 70% of maximum. The *V-port plug* at 50% opening delivers 35% of maximum flow, but beyond 50% opening it begins delivering flow at a steady but much higher rate. The *throttling plug* requires relatively large movement in the lower ranges of opening to affect flow. At 25% opening, only about 10% of flow is allowed. At 40% of opening there is 30% of maximum flow, and beyond that, flow increases rapidly with increased opening. Usually, in

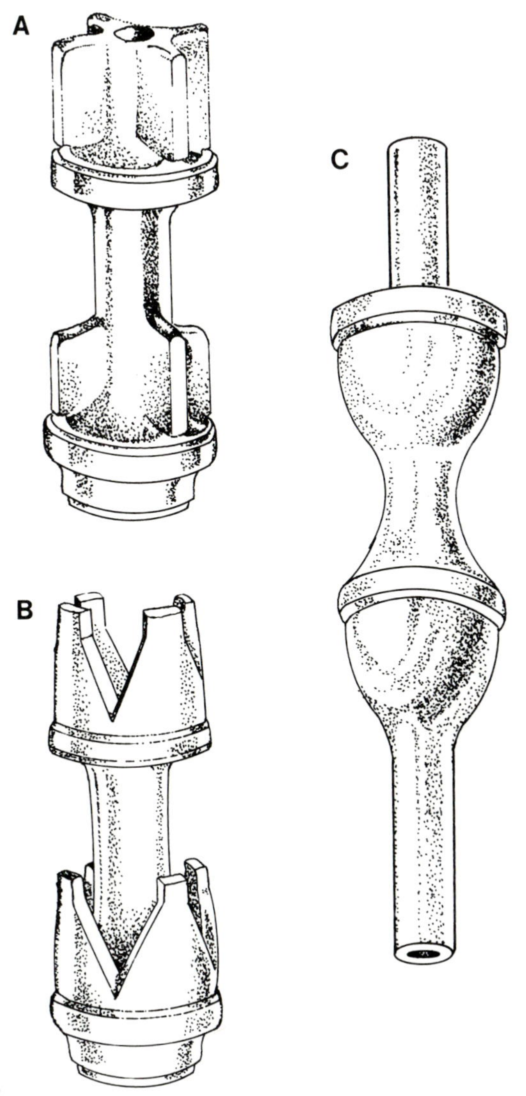

Figure 8.19. Types of valve plugs. *A*, quick-opening; *B*, V-port; *C*, throttling.

throttling service, flow rates through the valve are low, so this relationship of considerable valve plug movement to little change in flow rate is advantageous to good control.

Flow capacity describes the maximum volume of fluid that can flow through a wide open valve in a unit of time. To measure this capacity, two factors are taken into account: (1) pressure drop across the valve; and (2) viscosity, density, and type of fluid. A *flow coefficient* (*Cv*) is a convenient factor for expressing the flow characteristic of a valve. *Cv* is the volume of water in gallons that will flow through the wide-open valve in 1 minute with a pressure drop of 1 pound per square inch across the valve.

Valve actuators

A *valve actuator* (also called an operator) is a device that provides the force to move the valve stem and plug from one position to another. Actuators are classified according to the form of input signal and output power used. Thus actuators can be mechanical, pneumatic, electric, hydraulic, or a combination. For example, an electrohydraulic actuator is one that receives an electrical signal and uses hydraulic pressure to produce the mechanical motion of the valve stem.

Mechanical actuators. Mechanical actuators use a linkage to transmit motion between a sensing device and the valve stem. Such actuators are used in liquid-level control but are not sensitive enough to serve where accurate control is needed.

Pneumatic actuators. Pneumatic actuators are used extensively in the petroleum and chemical industries. They are the most popular actuators used in oil and gas field operations because of their safety, simplicity, reliability, and the generally ready availability of air or gas pressure for their operation.

The most common form of pneumatic actuator is one using a diaphragm against which air or gas pressure is applied (figs. 8.20 and 8.21). *Diaphragm actuators* usually contain a spring that opposes the pneumatic pressure applied against the diaphragm, although springless types, in which controlled pressure can be applied to either side of the diaphragm, are found in some services.

In a direct-acting spring-loaded actuator (fig. 8.20), pneumatic pressure is applied to the upper side of the diaphragm through an inlet at the top. Downward motion of the diaphragm and valve stem is opposed by the actuator spring. Downward motion tends to close a direct-acting valve and to open a reverse-acting one. Close inspection of figure

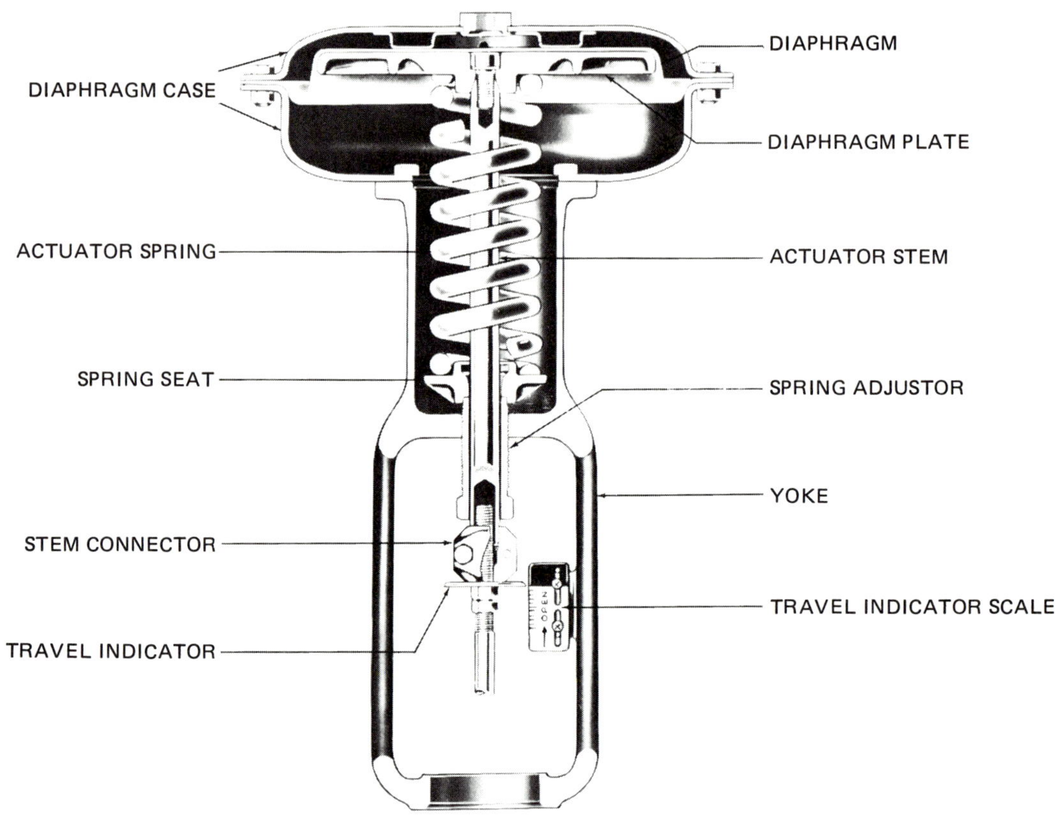

Figure 8.20. Spring-loaded, direct-acting diaphragm actuator (Courtesy of Fisher Controls)

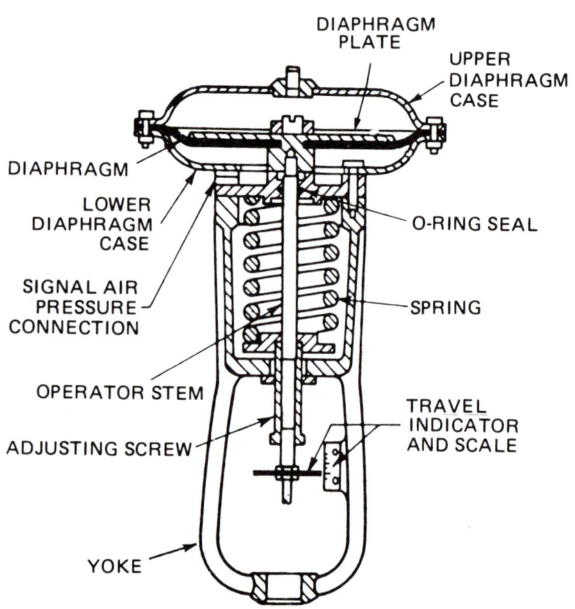

Figure 8.21. Spring-loaded, reverse-acting diaphragm actuator

8.21 shows that enough pressure applied to the lower section of the diaphragm case will drive the operator stem up, compressing the spring and opening a direct-acting valve.

Two forces are important in the operation of spring-loaded diaphragm actuators: (1) force caused by pneumatic pressure bearing on the diaphragm, and (2) force needed to compress the load spring. Force resulting from pneumatic pressure is the product of *area of the diaphragm* in square inches and *pressure* in pounds per square inch. Force for the spring is the product of *movement in inches* and a *factor k.* In acceptable abbreviations,

$$F = Ap = kx, \text{ or simply } Ap = kx$$

where

F = force in pounds
A = area in square inches

p = pressure in pounds per square inch
x = linear movement in inches
k = a constant factor for the particular spring in use.

If the spring is not in tension and rests against the diaphragm assembly with virtually zero force, the valve stem will begin to move once the gauge pressure on the diaphragm exceeds zero. Generally, actuators have the spring compressed a small amount, so the valve is firmly closed or open with zero pressure on the diaphragm. For this condition, the above equation is expanded to the following form:

$$Ap = kx + Fi$$

where

Fi is the initial force to be overcome before the valve stem moves.

Manufacturers and control system engineers are agreed on a standard range of pneumatic pressure that is in general use—3 to 15 pounds per square inch (psi). Thus when 3-psi pressure is applied to the actuator diaphragm, the valve stem will just begin to move, and at 15 psi applied pressure, the stem will reach full travel.

Suppose an actuator using the standard 3- to 15-psi range has total stem travel of 1 inch and a diaphragm area of 75 square inches. At 3-psi pressure, the stem is virtually "weightless," and at 15 psi it has traveled 1 inch.

For 3 psi:

$$3 \text{ psi} \times 75 \text{ in.}^2 = kx + Fi$$

However, there is no stem travel at this pressure ($x = 0$); therefore,

$$Fi = 3 \text{ psi} \times 75 \text{ in.}^2 = kx + 225 \text{ pounds.}$$

For 15 psi:

$$15 \text{ psi} \times 75 \text{ in.}^2 = kx + 225 \text{ pounds.}$$

Here, 225 pounds has been substituted for *Fi*. Stem travel is 1 inch ($x = 1$). The value of k can be determined by substituting 1 inch for x.

$$k \times 1 = 15 \text{ psi} \times 75 \text{ in.}^2 - 225 \text{ pounds}$$
$$k = 1125 - 225 = 900 \text{ pounds.}$$

The following equation can be used to determine the stem movement, x, in the above problem for any applied pressure between 3 and 15 psi:

$$x = \frac{p - 3 \text{ psi}}{12}$$

The equation indicates that the valve stem could assume any position in its range of travel by applying a predictable pressure to the diaphragm. In practice this might not be true, because friction, forces of flowing fluid, and other factors can cause effects that are not accounted for. Special devices called *valve positioners* are used to aid the actuators when very accurate positioning of the valve stem is required.

A *piston actuator* (fig. 8.22) may be designed to function with either pneumatic or hydraulic pressure. Piston actuators possess two advantages that make them desirable for many applications: (1) rugged design that allows handling high operating pressures, enabling them to provide quick response and great linear force; and (2) capability of delivering very large linear movement in comparison with diaphragm actuators.

In the actuators of figure 8.22 a valve positioner rides atop actuator *A*, while *B* is for on-off service. The on-off version requires that the cylinder be loaded and unloaded by a solenoid valve, pneumatic switching device, or similar equipment.

As noted earlier, a valve positioner is a device that aids the actuator in positioning the valve plug to an accurate setting. It accomplishes its purpose by acting as a force amplifier between the controller and the actuator, the controller being that part of the system that sends a pneumatic or other signal to the valve.

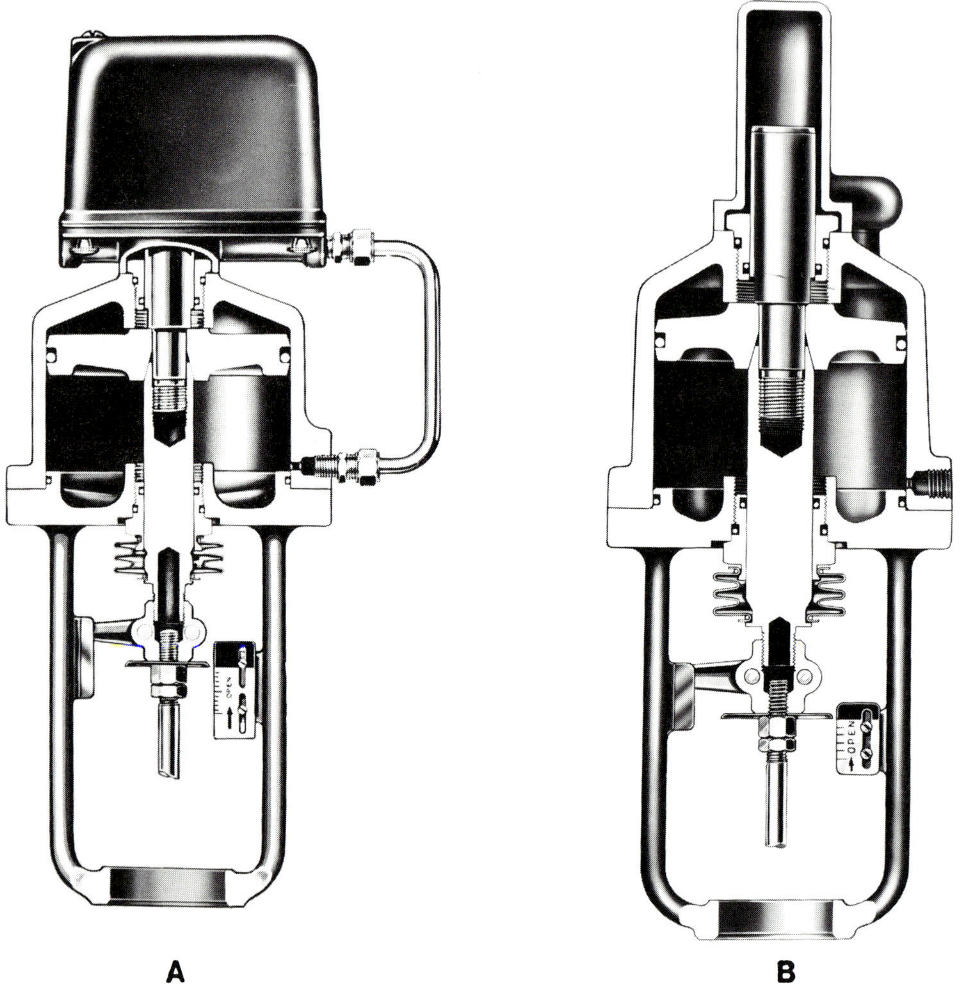

Figure 8.22. Pneumatic piston actuators. A valve positioner is attached to actuator *B*. (Courtesy of Fisher Controls)

Pneumatic control

Pneumatic control is the common method of sending signals and operating valve actuators in oil and gas fields; it is a form of control characterized by simplicity, reliability, and the ease with which it can be adapted to most situations. A reliable source of clean, dry air or natural gas at a pressure of 50 psi or so is all that is needed to power a complete system.

As noted in the section on valve actuators, a standard pressure range of 3 to 15 psi is needed to operate the usual valve actuator. This same range of values is also used to transmit information from one point to another in the control system. This information can pertain to values of temperature, pressure, flow, or liquid level. For transmitting information the pressure range of 3 to 15 psi is considered to be the analog of the range of values of the variables. A pressure of 3 psi might represent a temperature of 32°F, and 15 psi represent a temperature of 212°F. That is, for each 1 psi of pressure change, the temperature change will be (212°F − 32°F) divided by (15 psi − 3 psi) = 15°F/1 psi. This idea of analog values should become clearer as the study of pneumatic control continues.

Pressure regulators

The available pneumatic pressure at an oil or gas field is apt to be well in excess of that needed for pneumatic control. It is a good idea to study how pressure is reduced and regulated for control purposes. Generally, the pressure for a control system source will be 20 psi. A pressure regulator will be used to obtain that value.

A *spring-loaded regulator* of rugged but simple design is commonly used to regulate pressure for the control system (fig. 8.23). The compressed spring tends to drive the valve plug open, while the pressure existing at the outlet side of the unit exerts an opposing force against the lower side of the diaphragm. The valve plug will settle at some opening determined by the amount of pressure needed to balance the force of the compressed spring. Outlet pressure is adjusted by varying tension of the spring with the spring-compressing screw.

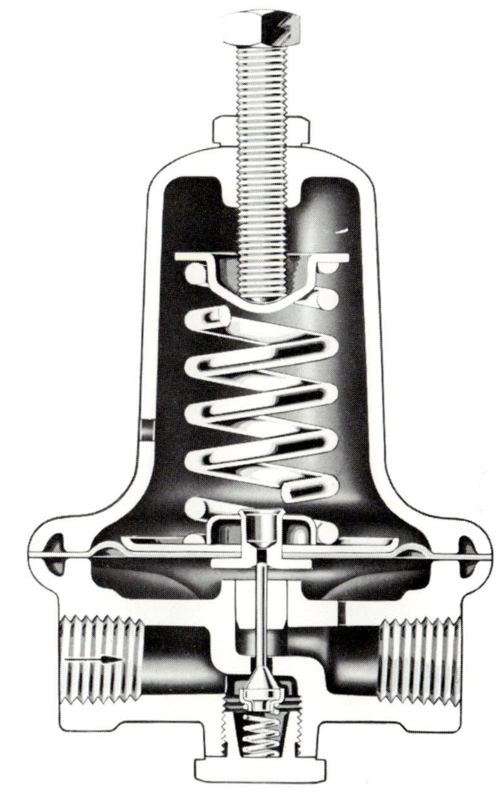

Figure 8.23. Self-contained, spring-loaded regulator (Courtesy of Fisher Controls)

A simple pneumatic control system

A pneumatic controller contains a *fixed* and a *variable* orifice (fig. 8.24). The two

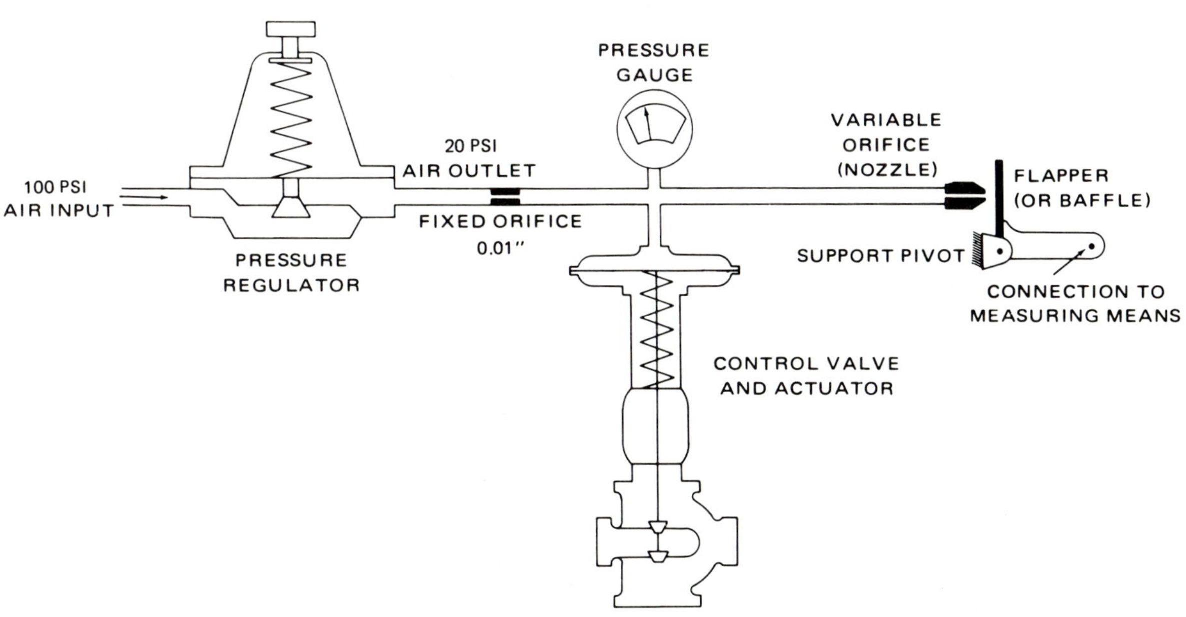

Figure 8.24. A simplified pneumatic controller

orifices are the basis for providing a means to modulate, or vary, the pressure over the standard range of 3 to 15 psi. In the diagram, pressure on the left side of the fixed orifice is maintained at 20 psi. The fixed orifice is small, being about 0.01 inch in diameter, and not capable of sustaining a high rate of fluid flow. The variable orifice is in the form of a nozzle, and has a flow diameter of about 0.025 inch. It is a variable orifice because a flapper or baffle is positioned to restrict flow from it. The flapper is connected to a measuring means, perhaps a float used in a liquid-level control system or the end of a Bourdon tube that is responding to pressure.

Should the flapper be moved away from the nozzle only a very short distance, say 0.01 inch, the air pressure between the fixed orifice and the nozzle will drop rapidly, because the fixed orifice cannot pass enough fluid to make up for the loss from the much larger orifice at the nozzle. Note that pressure between the orifices, commonly referred to as *nozzle back-pressure,* is applied to the diaphragm of the control valve actuator. In theory this system provides a means of controlling the pressure to the diaphragm actuator. The control is based on movement of the flapper, which in turn is controlled by the value of the controlled variable, whether it be temperature, pressure, or liquid level.

A practical pneumatic controller

In practice the simple system described above will be satisfactory for only a few applications. The orifices must be small to avoid excess waste of compressed air or gas; they will therefore seriously retard action of the diaphragm actuator. The diaphragm case is relatively huge and requires perhaps minutes to fully respond to the pressure changes existing between the two orifices. The system has other problems too, but all of them can be overcome to develop the basic idea into an effective controller.

Slow action of the diaphragm actuator can be overcome with an *air relay* (fig. 8.25). In the simplified version shown in the figure, air pressure between the two orifices is applied to

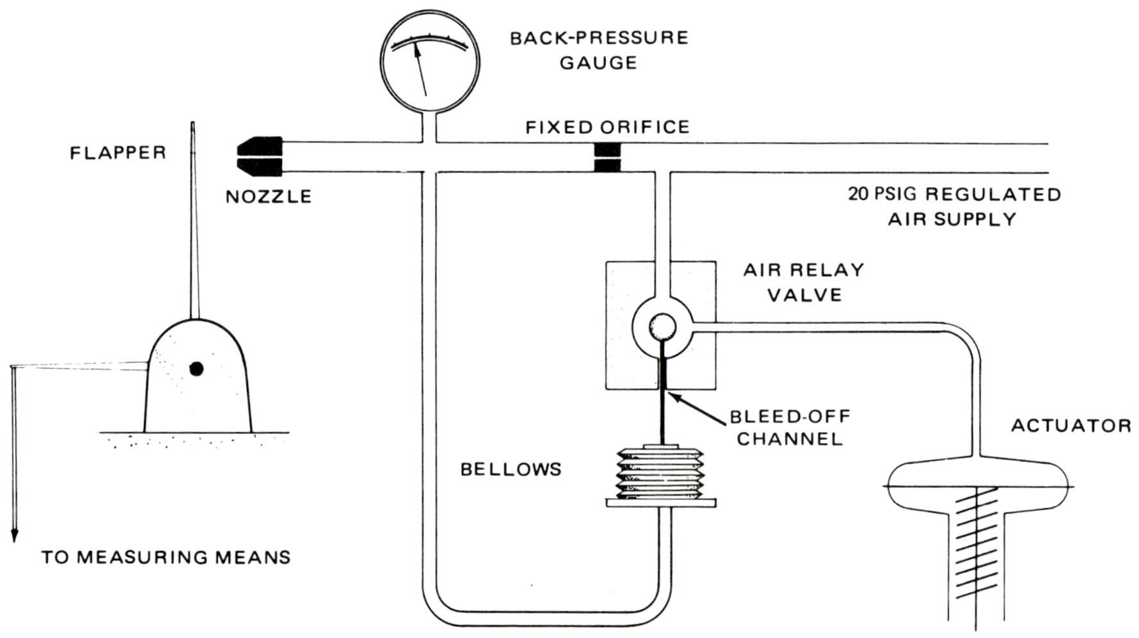

Figure 8.25. Use of an air relay to provide linear control

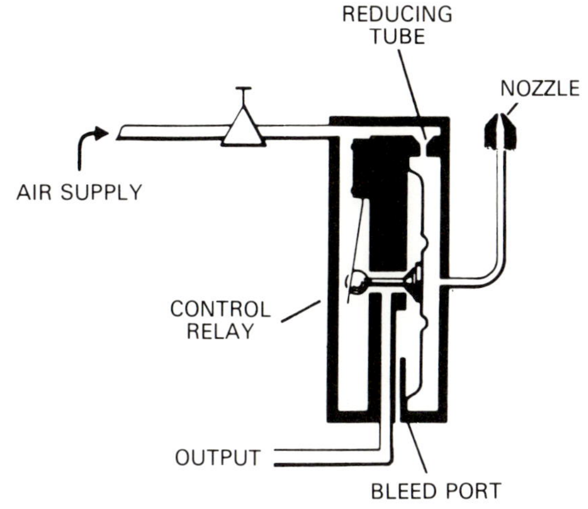

Figure 8.26. Continuous-bleed air relay (Courtesy of Foxboro)

the bellows of the air relay. This bellows has very little volume compared to the diaphragm actuator, so it responds rather rapidly to changes in nozzle back-pressure. The bellows drives a ball-and-seat arrangement that controls flow to the diaphragm actuator. This flow comes directly from the 20-psi pressure source, and the relay is capable of passing flow at a rate that provides satisfactory actuator response. The bellows contains a spring that can adjust output pressure to the limits of 3 to 15 psi. A commercial version of an air relay (fig. 8.26) uses a small diaphragm instead of a bellows to actuate interior valving. The fixed orifice is called a *reducing tube* in the illustration. This is a "continuous-bleed" relay, meaning that a very small amount of gas or air is continuously leaked from the device. When natural gas is used as the pressure agent, thought must be given to a possible hazard. However, for outdoor use this continuous-bleed feature poses no problem. Non-bleed relays are available. They are more complex, but just as reliable and sturdy.

A proportional controller

In the simple control systems considered so far the action has been very much a matter of *on* or *off*. The movement needed between the flapper and the nozzle to produce an output pressure of 3 to 15 psi, the full range of control pressure, is only about 0.01 inch. It is necessary to provide some method that will allow a movement of at least 0.25 inch for the measuring-means linkage. In fact, the method must also allow this quarter-inch of movement to move the flapper only 0.002 inch, because the really useful portion of the flapper's movement is that small, just 0.002 inch. As difficult as this situation might seem, it can be accomplished with good results.

Proportional controllers are in common use for many situations. When a proportional controller reacts to a change in the controlled variable, its response will be to change the manipulated variable in proportion to the change in the controlled variable. Say the energy output of a hot water system increases by 10%. The change will cause the controlled variable (temperature, in this case) to decline. The proportional controller will respond by admitting greater steam flow (the manipulated variable) to the heater. In fact, the energy input to the heater will rather quickly equal the energy output again, but this action will not cause the controlled variable (temperature) to go back to its original set point value, at least not until the demand at the output returns to normal. Despite this apparent shortcoming, proportional control serves well in situations where deviations above and below the set point are small and can be tolerated without harm.

A basic proportional controller is shown in figure 8.27. The illustration reveals how a rather large movement of the measuring-means linkage can be effected by using the extremely small movement of the measuring means. The flapper will be actuated by two variables: (1) action of the measuring means, and (2) the related action of the relay output pressure. The measuring means displaces the flapper. Then a bellows-spring arrangement operated by the air relay output pressure

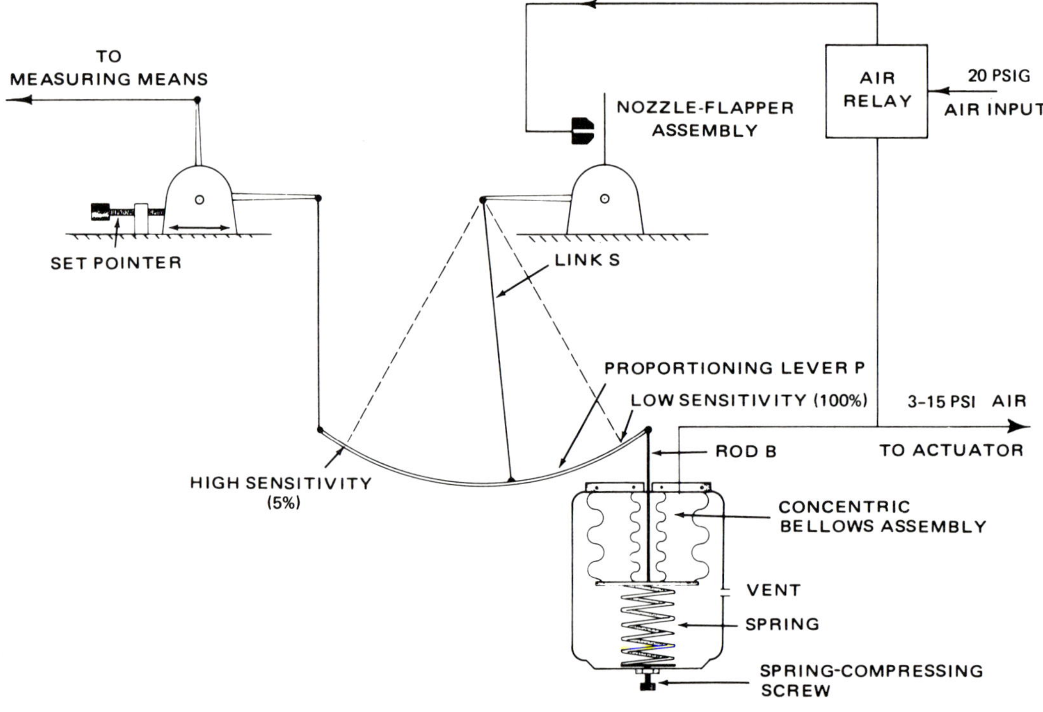

Figure 8.27. Proportional controller with a bellows spring assembly

drives the flapper back near its previous position. This action provides a form of negative feedback and adds stability to the system.

Rod *B* is a stiff member with vertical motion only. Proportioning lever *P* is a curved bar having a radius equal to the length of link *S*. Lever *P* is pivoted at the three points shown, but the pivot point formed with link *S* is adjustable from one end of lever *P* to the other.

The bellows-spring assembly consists of a pair of bellows arranged concentrically and fitted to common end plates. The bellows assembly is fitted into a sturdy metal cylinder in such a way as to compress a spring. A set screw adjusts tension of the spring. The unit comprises two airtight compartments, one of which is the bellows and the other the metal cylinder surrounding the bellows and spring; note that the cylinder is vented. Design is such that rod *B* will move a little more than 0.25 inch when pressure applied to the bellows is varied from 3 to 15 psi.

When regulated air pressure is applied to the control system, assuming the flapper is positioned well away from the nozzle, the nozzle back-pressure will be at a minimum. For this condition the air relay (fig. 8.25) will rapidly transmit air pressure to the valve actuator and to the bellows (fig. 8.27). As the bellows expands, rod *B* pulls down on lever *P*, causing the flapper to approach the nozzle. The resulting back-pressure between the nozzle and the fixed orifice quickly positions the ball-plug valve of the air relay. The system quickly reaches equilibrium.

Link *S* can be positioned at any point along lever *P*. With link *S* at the left end of lever *P*, movement of the measuring-means linkage will produce an almost equal movement of the flapper, while movement of rod *B* has little or no effect on flapper position. The controller will be extremely sensitive to changes in the controlled variable, because only about 0.002 inch of flapper movement can cause a full range of output (3 to 15 psi) from the air relay. This effect produces an on-off controller action.

Link *S* is now placed near the right end of lever *P*. In this position link *S* is not affected greatly by the measuring-means linkage, although for the position shown, full movement of the measuring-means linkage produces full effective movement of the flapper, 0.002 inch. The controller is said to have *100% throttling range,* which is to say that *full movement of the measuring means* produces *full movement of the valve stem,* and that 50% movement of the measuring means will produce 50% movement of the valve stem, and so on.

Somewhere along lever *P,* link *S* can be pivoted so that a full 0.25 inch movement of the measuring-means linkage moves the flapper 0.125 inch, and full movement of rod *B* moves the flapper the same amount, but in the opposite direction. For this position of link *S,* movement of the measuring-means linkage over only half its total travel produces a full range of output from the air relay. This set-up provides a 50% throttling range, since only half the range of the measuring means produces a full range of air relay output.

As the throttling range becomes smaller, the controller becomes more sensitive and can maintain the controlled variable closer to the set point. However, narrowing the throttling range has a limit. A point will be reached at which the controller causes the controlled variable to oscillate above and below the set point. This condition is generally undesirable.

Proportional-plus-reset controller

If it is desirable to have proportional control and yet necessary to have the controlled variable return by automatic means to the set point, the arrangement of figure 8.27 can be adapted to provide a means for accomplishing proportional control with an automatic reset feature.

A few additions to the proportional controller will provide reset action (fig. 8.28).

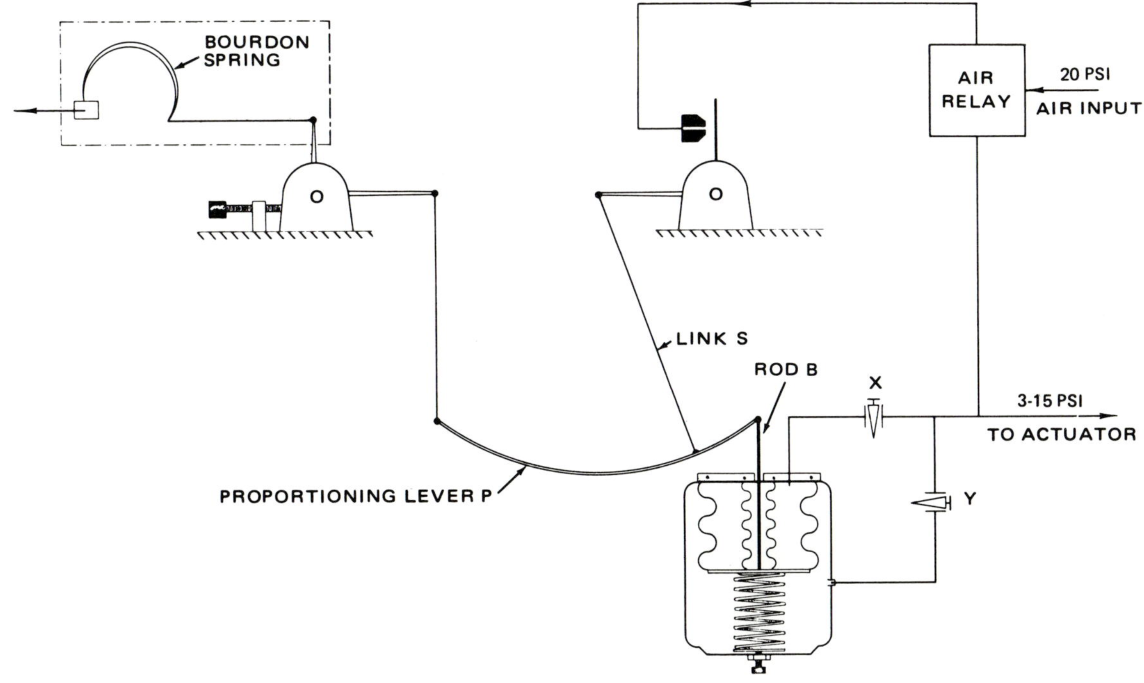

Figure 8.28. Proportional controller with automatic reset

Valves *X* and *Y* are needle valves. Valve *X* has been placed in the line from the air relay to the bellows. Valve *Y* is in a new line from the air relay to the cylinder surrounding the bellows. Should valves *X* and *Y* be wide open, rod *B* would be almost motionless for any output from the air relay, since equal pressure would exist on the interior and exterior of the bellows. Although the interior area is slightly smaller than the exterior area, the difference is not enough to matter. Operated in this manner, the controller will produce an on-off action.

Turning valve *Y* toward the closed position until the rate at which the cylinder becomes charged lags well behind the charging rate of the bellows will produce a desirable effect. Assume a 50% increase in demand for water from a hot water system. The measuring means detects a change in the temperature and causes the flapper-nozzle clearance to broaden. This action increases output from the reverse-acting air relay. A valve in the steam line to the heater will also be reverse-acting, so the increased pressure to the actuator will cause it to open the valve wider.

Increased pressure from the air relay will pass through valve *X* and expand the bellows unit. Rod *B* together with lever *P* and link *S* will tend to narrow flapper-nozzle clearance. The cylinder charges slowly, tending to keep the nozzle-flapper clearance broad. The reduced nozzle back-pressure ultimately results in wider opening of the steam valve, and this increased energy flow raises the temperature of the water. The system eventually settles to meet the new load, with temperature maintained at the set point. This is now a proportional-plus-reset controller.

Another controller mode, called *rate,* or derivative, is used in process control, but finds little use in oil and gas field operations. The rate mode is a feature that adjusts the rate of response of a control system, based on the speed at which the controlled variable is changing.

Field applications of instruments

To this point various methods and instruments for measurement and control have been studied. It is time to consider how these instruments and methods are applied in practice, and how they relate to field handling of natural gas. A logical place to start is the wellhead of a producing well.

The wellhead

The wellhead may contain an assortment of valves. Most valves are of the manually-operated through-conduit variety of gate valves (fig. 8.18). When fully open, these valves offer little impediment to flow of fluids. Chokes are needed to regulate flow. Most of the pressure drop between a well and field processing equipment occurs across the choke, so it is designed to better withstand the abrasive effects of high-pressure flow than typical throttling valves. Chokes may be adjustable (fig. 8.29) or fixed. The fixed choke is also referred to as a positive choke. Chokes commonly have right-angle flow, so they are not installed in a straight line with the wellbore.

Safety shutdown valves

Automatic shutdown valves form a part of the wellhead equipment on land wells. Well shut-in valves are installed as subsurface safety equipment for wells drilled on offshore locations.

A proper safety shutdown valve (SSV) is a fail-safe device, meaning that it will shut in the well in the event of overpressure or underpressure downstream of the well, or if the actuating force for the valve fails. The valve may be installed in line with the wellbore, or located in a line leading at right angles to the wellbore. It will be a through-conduit gate valve.

The need for automatic shutdown arises from the possibility of overpressure occurring

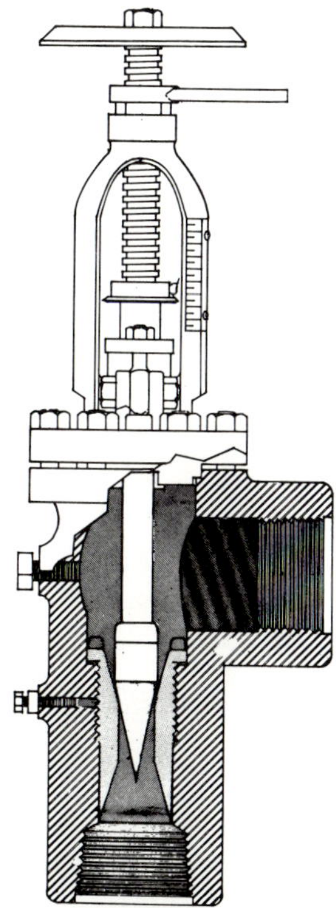

Figure 8.29. Adjustable choke

due to hydrates, the improper closing of a flow-line valve, or any other event that blocks the downstream flow from the wellhead. The choke in the line serves to control flowing pressure downstream, but it will not prevent full wellhead pressure from existing downstream if normal flow is stopped. Ordinary field treating or processing equipment is not likely to have the ability to withstand wellhead pressure, so it must be protected. Sudden loss of pressure downstream of a choke can be the result of a ruptured flow line or failure of field treating equipment, so it is important to shut off flow to prevent the loss of well production or something worse.

A fail-safe shut-in valve at the wellhead will usually be a spring-loaded, pneumatically-actuated valve using air or gas as the pneumatic pressure source. The actuator will be of the piston variety because it will have the linear travel necessary to position the gate plug. The load spring acts to close the valve in the event of loss of pneumatic pressure to the piston.

Pneumatic pressure applied to the SSV will compress the load spring and open the valve. Loss of this pneumatic pressure for any reason will cause the valve to close. Pneumatic pressure to the actuator relies on actions of three elements – two pressure sensors and a relay. One of the pressure sensors will react to low pressure, the other to high pressure. The two sensors are identical in construction, but their placement in the system and their adjustment enable them to serve either position. Figure 8.30 is a detailed drawing of their construction. The relay is simply a pneumatic

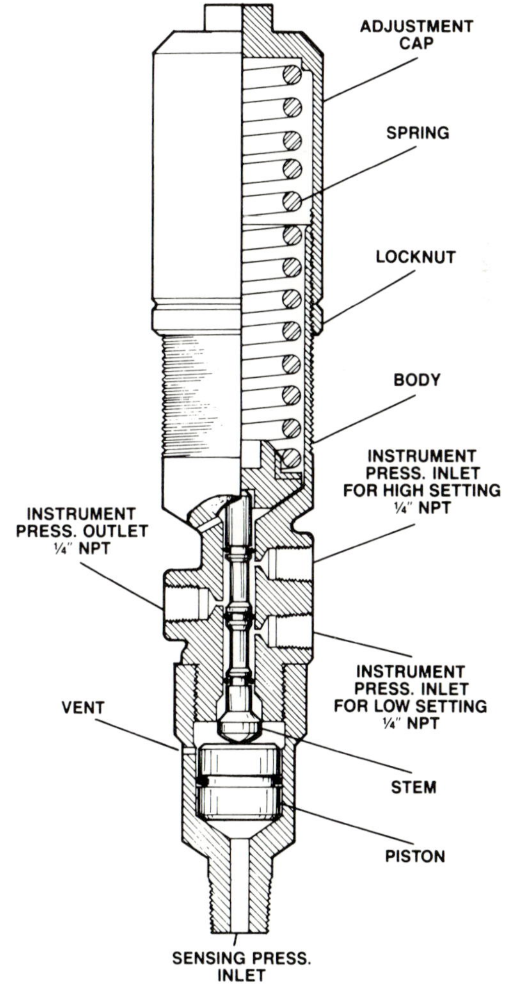

Figure 8.30. Pressure sensor designed to respond to limits of either high pressure or low pressure

SURFACE SAFETY SYSTEM   NORMAL OPERATION

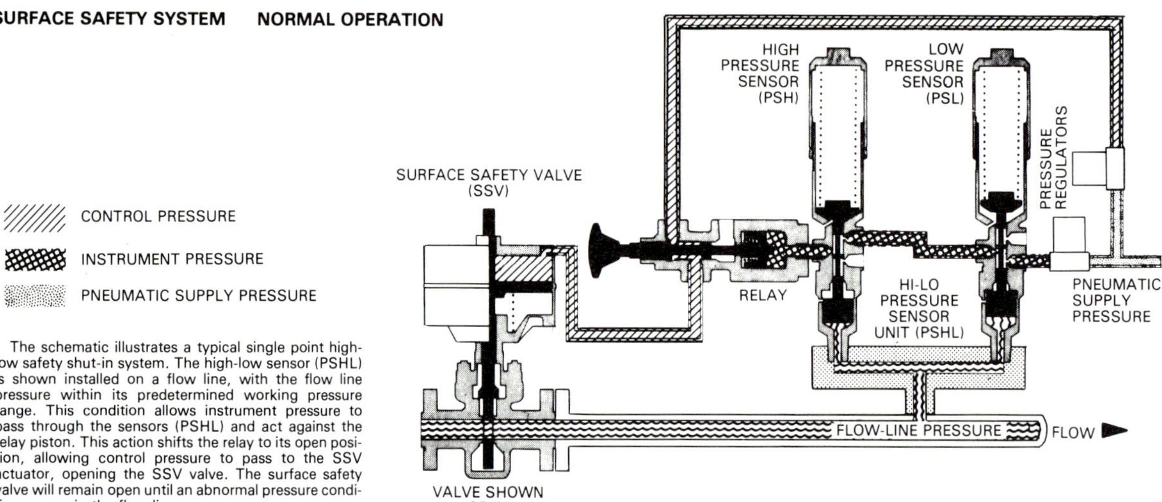

CONTROL PRESSURE

INSTRUMENT PRESSURE

PNEUMATIC SUPPLY PRESSURE

The schematic illustrates a typical single point high-low safety shut-in system. The high-low sensor (PSHL) is shown installed on a flow line, with the flow line pressure within its predetermined working pressure range. This condition allows instrument pressure to pass through the sensors (PSHL) and act against the relay piston. This action shifts the relay to its open position, allowing control pressure to pass to the SSV actuator, opening the SSV valve. The surface safety valve will remain open until an abnormal pressure condition occurs in the flow line.

SURFACE SAFETY SYSTEM   HIGH-PRESSURE SHUT-IN

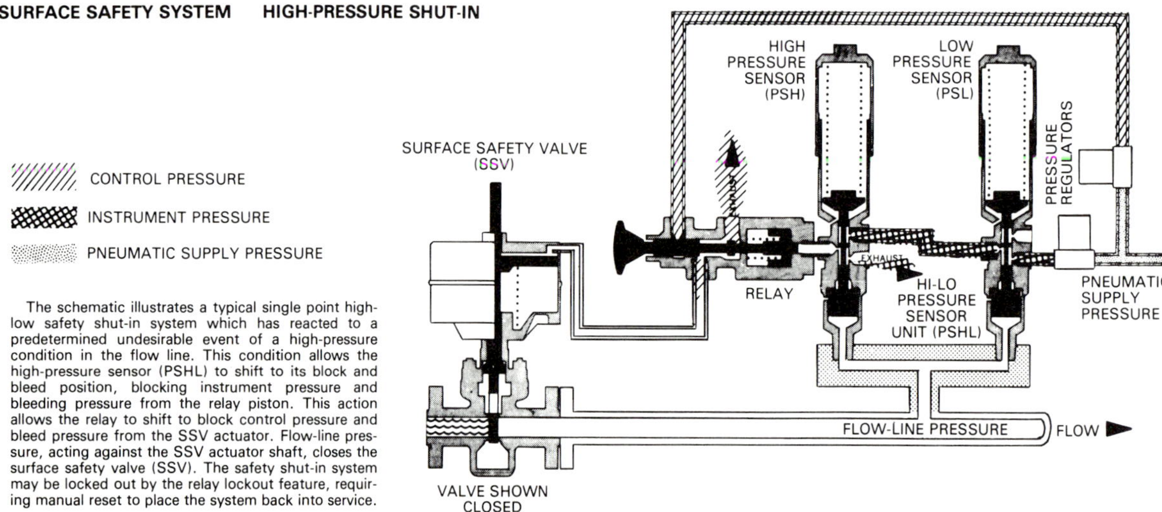

CONTROL PRESSURE

INSTRUMENT PRESSURE

PNEUMATIC SUPPLY PRESSURE

The schematic illustrates a typical single point high-low safety shut-in system which has reacted to a predetermined undesirable event of a high-pressure condition in the flow line. This condition allows the high-pressure sensor (PSHL) to shift to its block and bleed position, blocking instrument pressure and bleeding pressure from the relay piston. This action allows the relay to shift to block control pressure and bleed pressure from the SSV actuator. Flow-line pressure, acting against the SSV actuator shaft, closes the surface safety valve (SSV). The safety shut-in system may be locked out by the relay lockout feature, requiring manual reset to place the system back into service.

SURFACE SAFETY SYSTEM   LOW-PRESSURE SHUT-IN

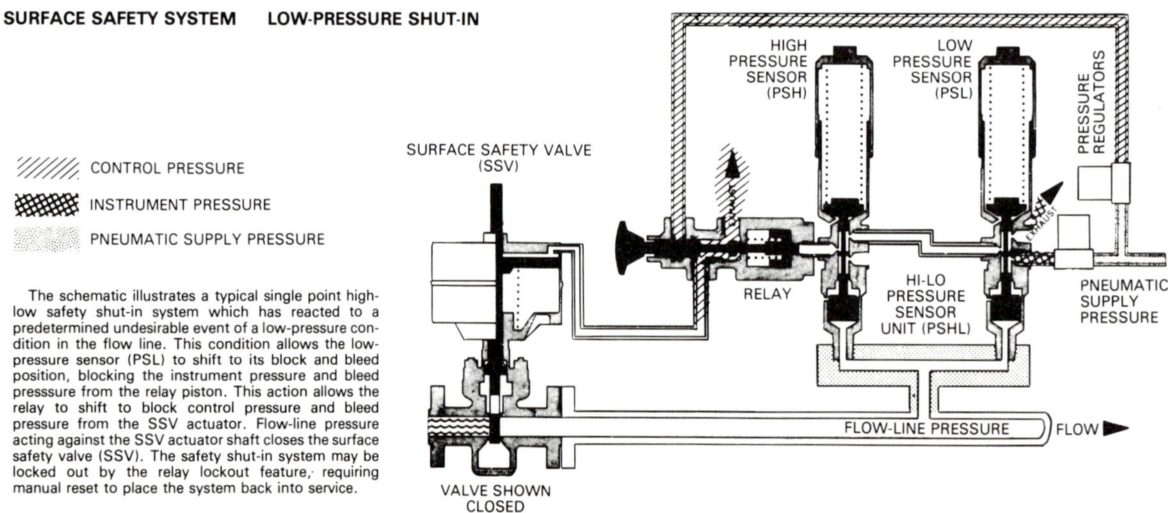

CONTROL PRESSURE

INSTRUMENT PRESSURE

PNEUMATIC SUPPLY PRESSURE

The schematic illustrates a typical single point high-low safety shut-in system which has reacted to a predetermined undesirable event of a low-pressure condition in the flow line. This condition allows the low-pressure sensor (PSL) to shift to its block and bleed position, blocking the instrument pressure and bleed presssure from the relay piston. This action allows the relay to shift to block control pressure and bleed pressure from the SSV actuator. Flow-line pressure acting against the SSV actuator shaft closes the surface safety valve (SSV). The safety shut-in system may be locked out by the relay lockout feature, requiring manual reset to place the system back into service.

Figure 8.31. The three stages in the operation of pressure sensors, relay, and surface safety valve

device with a spring-weighted piston driving a sliding valve. With pressure on the piston, the sliding valve is open to allow pneumatic pressure to be applied to the motor mechanism of the SSV. This keeps the valve open (fig. 8.31*A*).

In figure 8.31*B* the flow-line pressure has exceeded the maximum for which the system is adjusted. The piston of the high-pressure sensor (*PSH*) overcomes tension of the adjusting spring. This action simultaneously closes off air pressure from the pneumatic supply and opens a path to vent the relay. The relay in turn shuts off air to the SSV actuator and provides a path for venting pressure from it.

In figure 8.31*C* the flow-line pressure has fallen below prescribed value, causing spring tension in the low-pressure sensor (*PSL*) to move the stem downward. This action closes off the pneumatic supply to, and provides a vent path for, the relay through the *PSH* and out the *PSL*. The relay vents the valve actuator, and the valve closes.

The high-pressure and low-pressure sensors may be placed anywhere downstream of the wellhead. A typical location is at the entry to the first stage of treating or separation equipment (fig. 8.32).

Once downstream conditions that caused the shutdown are corrected, the well can be restored to production by pulling on the knob or handle of the relay. Assuming pneumatic pressure to be present, this will provide a path for pressure to actuate the SSV.

Hydrocarbon separation units

Hydrocarbon separators are almost certain to contain instruments for controlling pressure, temperature, liquid level, and flow rate.

Figure 8.32. Pressure sensors, for control of the surface safety valve, located on a hydrocarbon separation unit

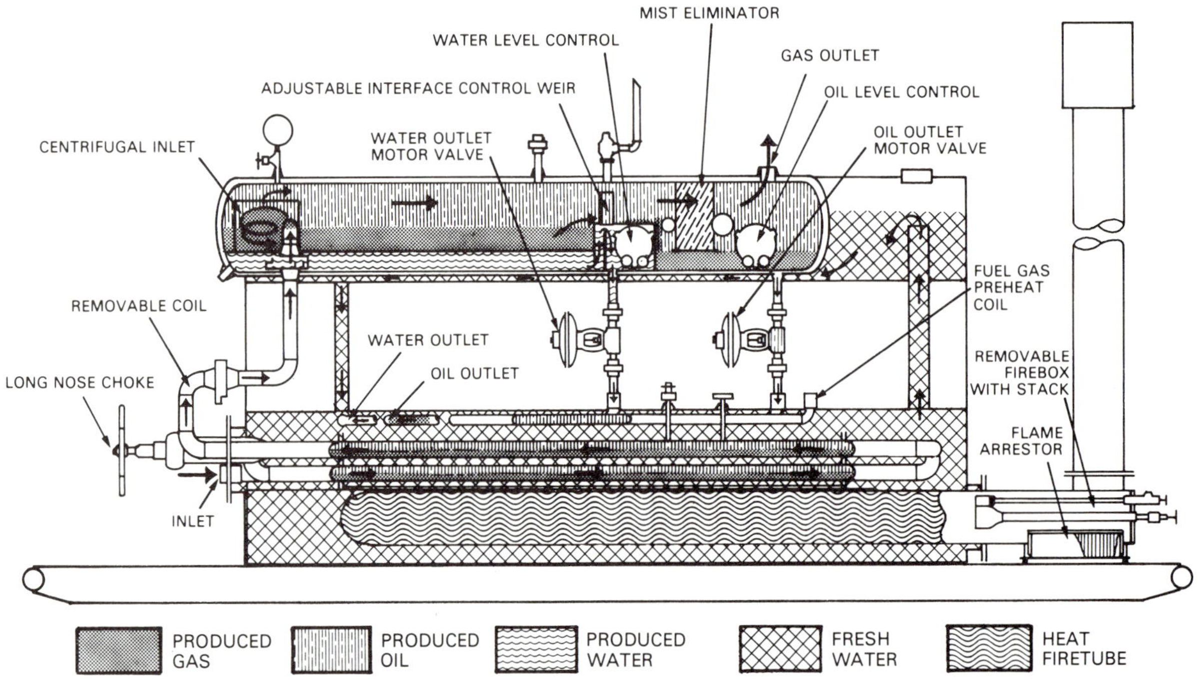

Figure 8.33. Stacked separation unit showing path of fluids and various control elements

Figure 8.34. A stacked unit similar to the one in figure 8.33. Liquids are driven up and out by gas pressure existing on their surfaces.

The flow line from the wellhead enters the separation unit where it is probably heated to some prescribed temperature (fig. 8.33). Fluids flow to the upper stage where water, condensate, and gas are separated. Water and oil or condensate are drained from the upper vessel as they reach appropriate levels in the lower part of the vessel. These liquids are "dumped" by valves that respond to liquid-level controllers. Gas passes out the top of the vessel, and its flow is controlled by a back-pressure controller. A thermostat controls temperature of the gas-fired heater section.

Liquid-level controls are on-off devices. Water and oil or condensate levels in a separator are held within limits by pneumatic motor valves (fig. 8.34). Floats in the water and oil section of the separator actuate pneumatic switches that allow pressure to be applied to the diaphragm when the level reaches an upper limit. At the lower limit the switch shuts off pressure to the diaphragm and provides a path to vent the diaphragm case.

The pneumatic switch may take the form shown in figure 8.35. A float in the vessel rotates the milled wheel through an arc of perhaps 30° or less. The wheel contains two radial arms that operate the switch handle, or toggle. In one position the switch sends pressure to the motor valve. In the other position it vents the valve actuator. The arms are easily adjusted to provide a differential gap that will give a wide range of level control.

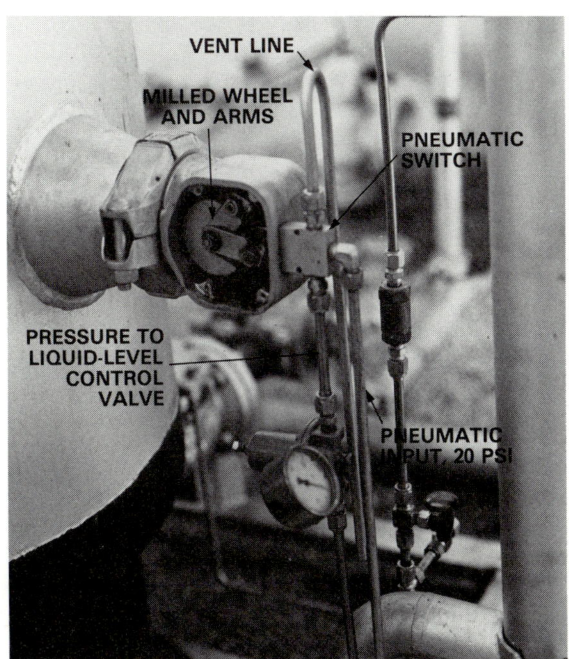

Figure 8.35. A pneumatic switch and its actuating mechanism. Arms on the milled wheel move the switch toggle from *on* to *vent*.

An effective method of controlling liquid level in a pressurized vessel resembles, in principle, the measurement device discussed in connection with figure 8.15. A diaphragm motor valve is actuated by pressure from a pilot assembly (fig. 8.36). The pressure is the differential between that due to height of liquid in the vessel and the gas pressure above the liquid.

The diaphragm valve uses discharge pressure from the vessel to force the plug away from its seat, while the diaphragm actuator drives the valve closed.

Gas pressure existing above the liquid in the vessel is applied to the top side of the pilot diaphragm, while pressure due to the combined effect of the hydrostatic head of liquid in the vessel and the gas pressure at the surface of the vessel liquid is applied to the lower side of the diaphragm. The pilot spring exerts a downward force on the diaphragm.

The pilot plug is a miniature barbell assembly. The upper ball controls flow of gas from the upper diaphragm chamber to the motor valve actuator and to the pilot spring chamber, while the lower ball controls venting of the motor valve actuator. Pressure existing in the chamber between the two balls exerts an upward force on the internal pilot assembly. In order to neutralize this force, the same pressure is also applied to the pilot spring chamber, where it may exert an equal downward force.

Rising liquid level in the vessel causes the pilot diaphragm assembly to be driven against the pilot spring. The upper ball will prevent flow of gas to the valve actuator, while the lower ball will lift from its seat, allowing the motor valve diaphragm pressure to bleed off.

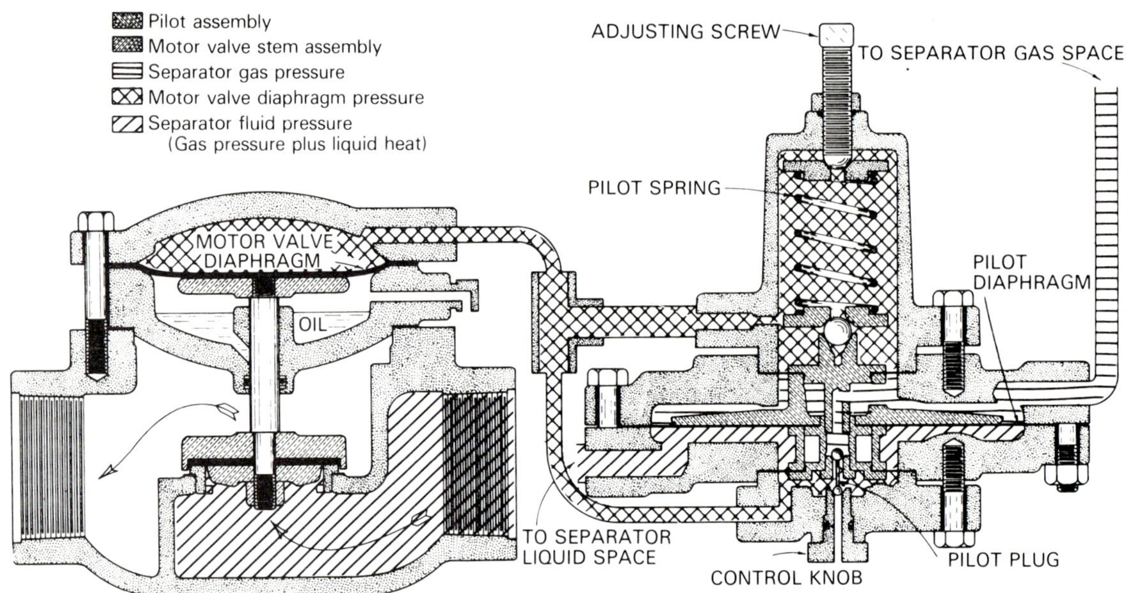

Figure 8.36. Diaphragm valve controlled by a pilot valve used to regulate liquid level (Courtesy of Kimray)

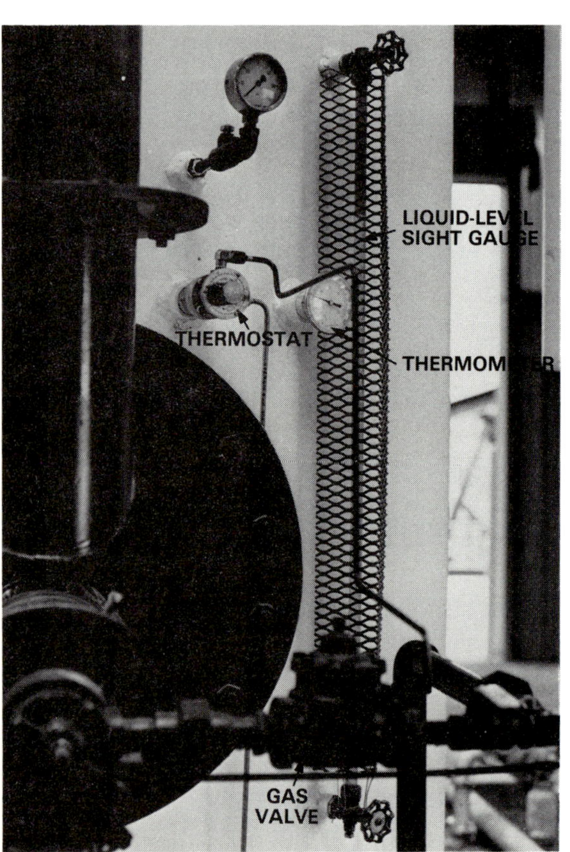

Figure 8.37. Thermostat, thermometer, pressure gauge, liquid-level sight gauge, and gas valve on a vertical heater-treater

Simultaneously, pressure will bleed off from the pilot spring chamber. Reduced pressure in the valve actuator will permit pressure from the vessel to force the valve plug away from its seat.

As the liquid level in the pressurized vessel decreases, so does the force acting on the lower side of the pilot diaphragm. Eventually the gas pressure and the pilot spring force the diaphragm assembly lower, restoring equilibrium.

Temperature control

Control of temperature in field treating equipment is largely a matter of controlling flow to a gas burner. A linear expansion type thermostat and thermometer are common instruments on heater treaters (fig. 8.37). The thermostat and gas valve (fig. 8.38) provide an on-off flow of gas to the burner. The thermostat contains a low-expansion alloy and a stainless steel tube as the temperature-sensitive element. With rising temperature the stainless steel tube increases in length to move the diaphragm assembly in a direction to close the seat at ball 1, and open the seat at ball 2.

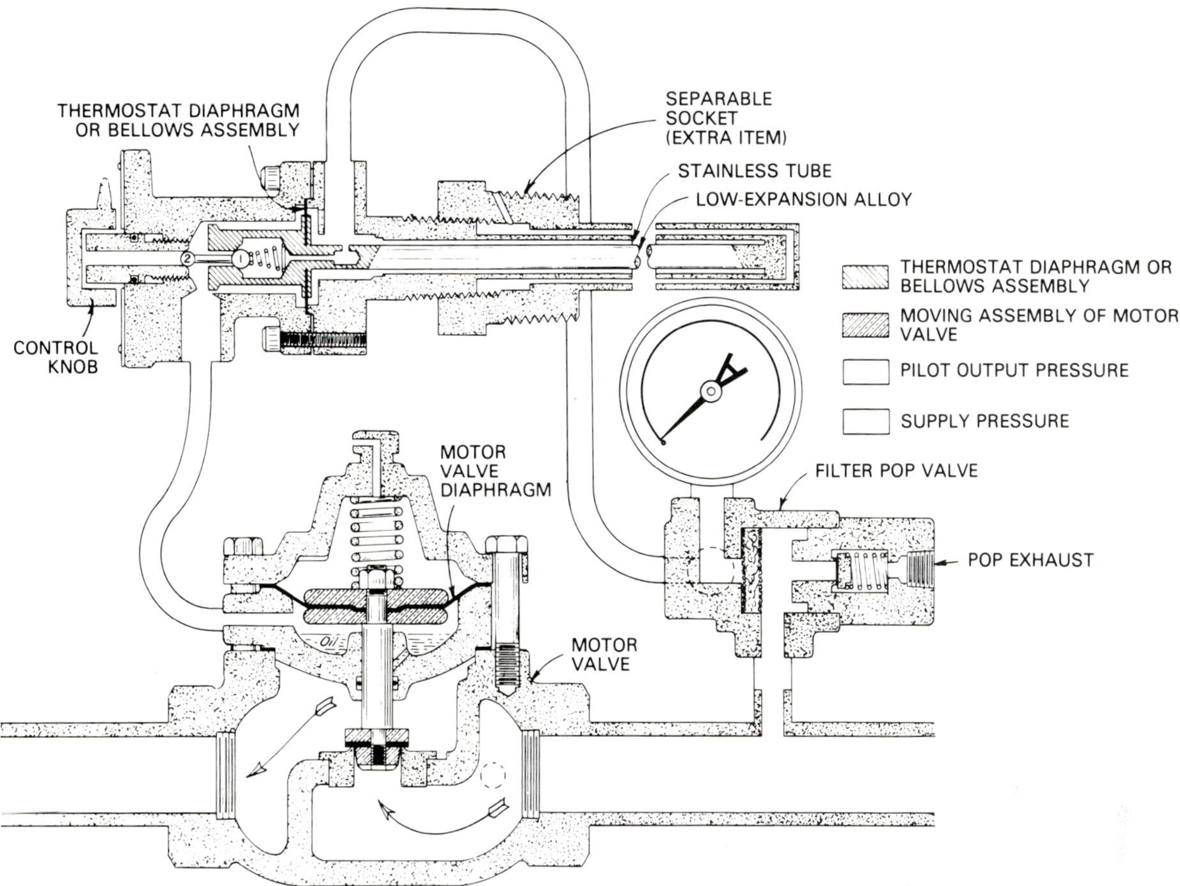

Figure 8.38. Combination thermostat and pneumatic motor valve used to control temperature in the heater-treater of figure 8.37

Opening the seat at ball 2 allows the diaphragm cavity of the motor valve to be vented, and the valve closes, shutting off the gas burner.

As temperature decreases, the above action reverses. Ball 2 will seat and ball 1 will unseat to allow pressure to the motor valve diaphragm.

Back-pressure control

It is generally necessary to control the pressure in oil and gas field units used for treating and separating the fluids. Such control is typically achieved by controlling the *flow* from the particular unit. A pneumatic proportional controller (fig. 8.39) mounted directly to the back-pressure valve is in common use. The proportional mode action is fully capable of the demands of control in this instance.

Figure 8.39. Motor valve controlled by a proportional controller to regulate back-pressure in a hydrocarbon separation unit

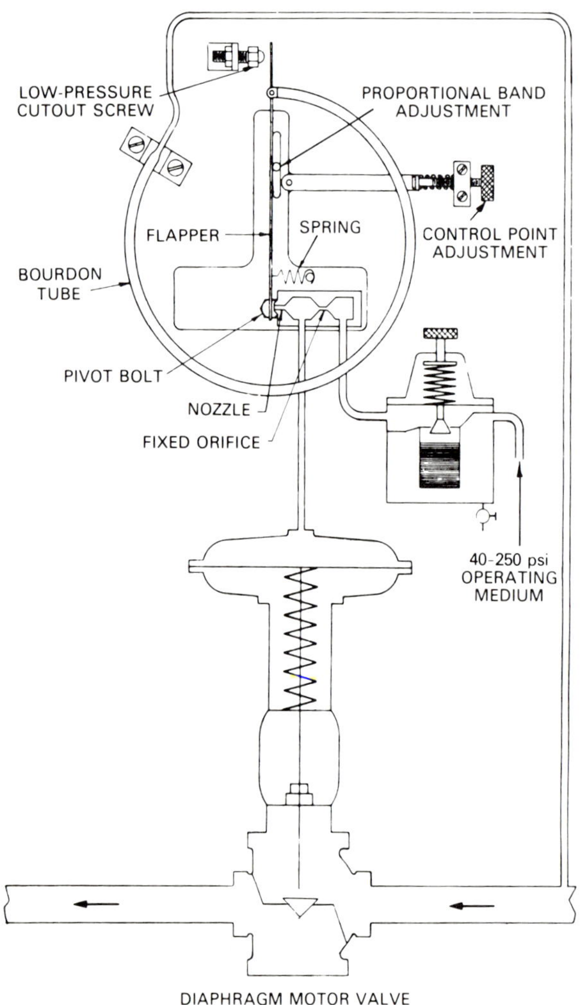

Figure 8.40. Diagram of a proportional band pressure controller

No air relay is used between controller and valve mechanism, but the response action between controller and valve is fast enough, since there are no lengthy pneumatic lines to retard action. The flapper is attached to the free end of the Bourdon tube and rides against an adjustable fulcrum pin, which is the proportional band adjustment (fig. 8.40). The lower part of the flapper is positioned over the nozzle, and a spring keeps the flapper against the proportional band adjusting pin. Tension of this spring prevents backlash that could arise from looseness between flapper and Bourdon tube. The control point adjustment moves the base plate assembly, including the flapper. Upper end of the flapper is attached to the Bourdon tube, which does not move during this adjustment, so the flapper is forced to pivot about the proportional band adjustment and thus change flapper-nozzle clearance. This change in clearance effectively alters the control point setting.

9 Measurement of natural gas and gas liquids

Natural gas is a valuable asset and a superior fuel. Good business practices dictate the need for efficient handling of natural gas and natural gas liquids. Efficient handling of these products includes good measurement practices.

The volume of gas is the fundamental basis for settlement in most gas-sales transactions. Payments for royalties and taxes are usually based on measured volumes. Reports to state regulatory bodies and to the Federal Power Commission, when required, are based on volumes measured in the field. Gas, being a vapor, is not subject to conventional methods of storage in large quantities. Therefore, it must be measured instantaneously as it flows through a pipeline. Natural gas liquids may be measured as they flow through a line, or they may be stored and measured in the storage vessel.

Metering of fluids may be described as the practice of measuring volumes or rates of flow by actually passing the fluids through some type of meter. While metering is the most common method of determining volumes or rates of flow, other methods of measurement are sometimes used. For gas, these include calculation of the volume in a pipeline (i.e., line pack) or in storage on the basis of the line or storage volume and pressure; application of pipeline flow formulas; estimation from time and rate of usage factors; computation of compressed volumes from temperature, pressure, cylinder displacement, and compressor speed; and other similar techniques that are practical. For liquids, the gauging of storage tanks would be included.

A fluid flowing through a line can be measured by placing a constriction in the line to cause the pressure of the flowing fluid to drop as it passes the constriction. This pressure drop is called *differential pressure* (fig. 9.1). A direct relationship exists between the rate of flow and the amount of this pressure drop, or differential. This principle has been widely used and has been developed into a precise and accurate means of measuring fluids when all factors are taken into consideration and when ideal conditions for its use prevail. Because this principle is so widely used, the method that makes use of the principle known as *orifice measurement* will be given emphasis herein.

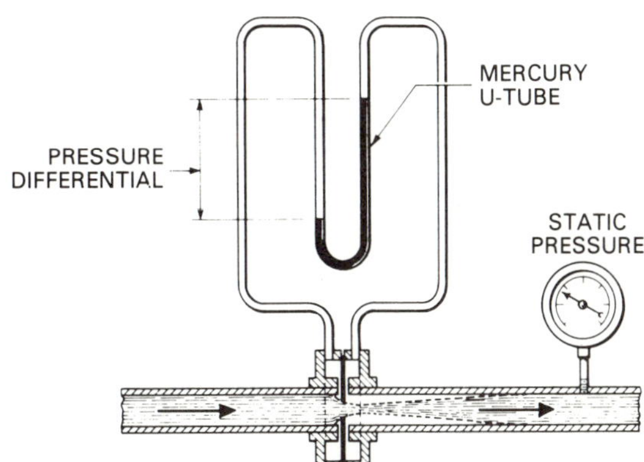

Figure 9.1. Diagram of the pressure drop across an orifice plate

Both gas and liquids may be measured by use of other measurement techniques including positive displacement meters, turbine meters, venturi meters, flow nozzles, critical flow provers, elbow meters, variable area meters (rotameters), and others. The selection of the measurement method to be used should be made only after careful analysis of several factors including the following:

1. accuracy desired;
2. expected useful life of the measuring device;
3. range of flow and temperature;
4. maintenance requirements;
5. power availability, if required;

6. whether liquid or gas is to be measured;
7. cost of operation;
8. initial cost;
9. availability of parts;
10. acceptability by others involved;
11. purpose for which measurements are to be used; and
12. susceptibility to theft or vandalism.

Natural gas measurement

Measurement of gas requires the unit of volume to be defined. In other words, the unit, temperature base, pressure base, and other factors must be specified or determined. There are currently no universally accepted temperature and pressure bases for cubic foot measurements although there may be in the future. The bases most likely to find acceptance are 60°F (520°R) and 14.73 psia. A cubic foot of gas may be defined as being the amount of gas required to fill a cubic foot of space when it is at a temperature of 60°F and under a pressure of 14.73 psia. Gas may be measured by an orifice meter in other units such as the pound, therm, MMBtu, and so forth. These, too, must be defined.

Orifice meters

The orifice meter consists of several parts. These parts make up what are usually referred to as the primary element (i.e., meter tube, etc.) and the secondary element (the recorder). Since gas is a highly compressible fluid, its density will vary considerably with the pressure existing at the constriction, or orifice plate. Consequently, when gas is measured with an orifice meter, it is necessary to measure both the differential pressure and the flowing pressure, which in gas measurement language is called the static pressure. There are several other measurements and factors to be taken into account in the accurate measurement of gas, such as temperature and specific gravity.

Primary element

The meter run, meter fittings or flange unions, and the orifice plate, which make up the primary element of a meter setting, should be a shop-fabricated unit manufactured by a reputable manufacturer of such equipment. The unit should be manufactured in accordance with the latest edition or revision of the American Gas Association's Gas Measurement Committee Report No. 3. The recommended standards on gas measurement expressed in Report No. 3 are the result of several years of intensive research and experimentation by special joint committees of the American Gas Association (AGA) and the American Society of Mechanical Engineers (ASME). The original report was approved for publication by the director of the National Bureau of Standards. Every revision has resulted from data collected during careful testing, and each revision has served to improve the utility of the recommendations published.

When a primary element is constructed in accordance with Report No. 3, the flow of gas through it will be streamlined, and there will be no obstructions within to disturb the flow pattern and contribute to inaccurate measurement. It is advisable to make direct reference to AGA Report No. 3 when planning or constructing a meter run. Report No. 3 and its specifications and recommendations relate to and should be understood to be limited to the following two types of orifice meters:

1. orifice meters with circular orifices placed concentrically in the pipeline and having upstream and downstream pressure taps as specified for flange taps; and
2. orifice meters with circular orifices located concentrically in the pipeline and having upstream and downstream pressure taps as specified for pipe taps.

Report No. 3 covers the measurement of flowing fluids, particularly natural gas, as related to the primary element and to the methods of calculation. The report does not cover equipment used in the determination of the pressures, temperatures, and other variables that must be known for accurate measurement. The installation sketch in figure 9.2 shows a basic meter tube length. This

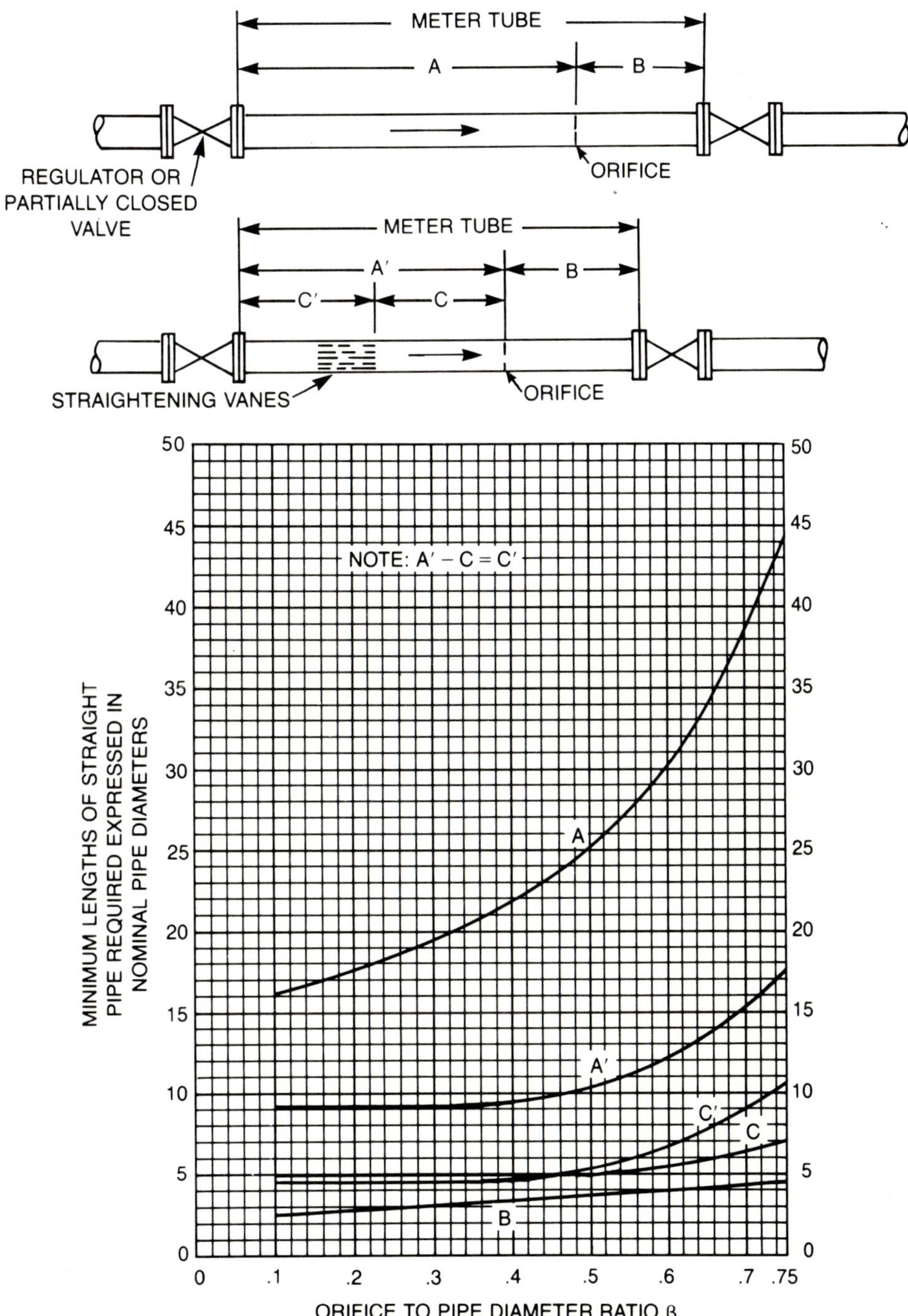

Notes: 1. When pipe taps are used, lengths A, A', and C shall be increased by two pipe diameters and B by eight pipe diameters.
2. When the diameter of the orifice may require changing to meet different conditions, the lengths of straight pipe should be those required for the maximum orifice-to-pipe diameter ratio that may be used.

Figure 9.2. Installation sketch for a basic orifice meter tube, from AGA Report No. 3, ANSI/API 2530, Second Edition

length of meter tube will accommodate a restriction in the pipeline upstream of the orifice. Meter tube lengths may be reduced under specific circumstances.

The many experiments made in development of the orifice flow constants and other factors given in AGA Report No. 3 involved the use of long, straight lengths of pipe both upstream and downstream of the orifice. Many of these experiments involved the use of straightening vanes. These are usually bundles of small-diameter tubing tack-welded together in a concentric pattern (fig. 9.3) and placed in the upstream section of the meter run, as indicated previously. Since most field applications at some time involve conditions reflected in the installation sketch shown in figure 9.2, it is recommended that field installations meet the requirements portrayed as to upstream and downstream lengths as well as to the use of straightening vanes. When space or other factors make it desirable to shorten upstream or downstream lengths, installation sketches shown in the AGA Report should be referred to.

The installation of straightening vanes will reduce considerably the amount of straight pipe required upstream of the orifice. Vanes should be reamed as thin as practical at both ends and should be arranged so that they are symmetrical with the smallest diameters in the center of the bundle, though the tubes need not be the same size. The purpose of the vanes is to eliminate swirls and crosscurrents set up by the pipe fittings and valves preceding the meter tube.

The orifice plate is held in position in the meter run by conventional orifice flanges or by a special orifice fitting. Orifice fittings are constructed to permit easy changing and inspection of orifice plates. Since such inspections where conventional flanges are used may be quite difficult, fittings are now generally used. Some fittings are designed so that plates can be changed without shutting off the flow of gas (fig. 9.4); in others, the gas must be shut off or the meter run bypassed. The circumstances peculiar to a given measuring situation govern the decision as to the type of installation to be made.

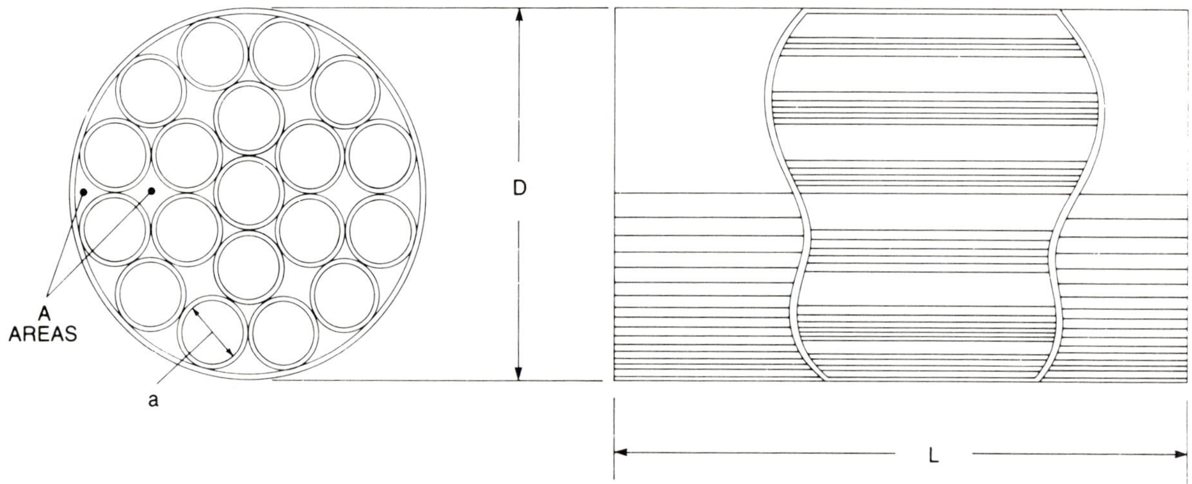

A: Cross-sectional area of any passage within the assembled vanes

a: Maximum transverse dimension of any passage through the vanes

D: Inside diameter of meter tube

L: Length of the vanes

Figure 9.3. Straightening vanes, from AGA Report No. 3, ANSI/API 2530, Second Edition

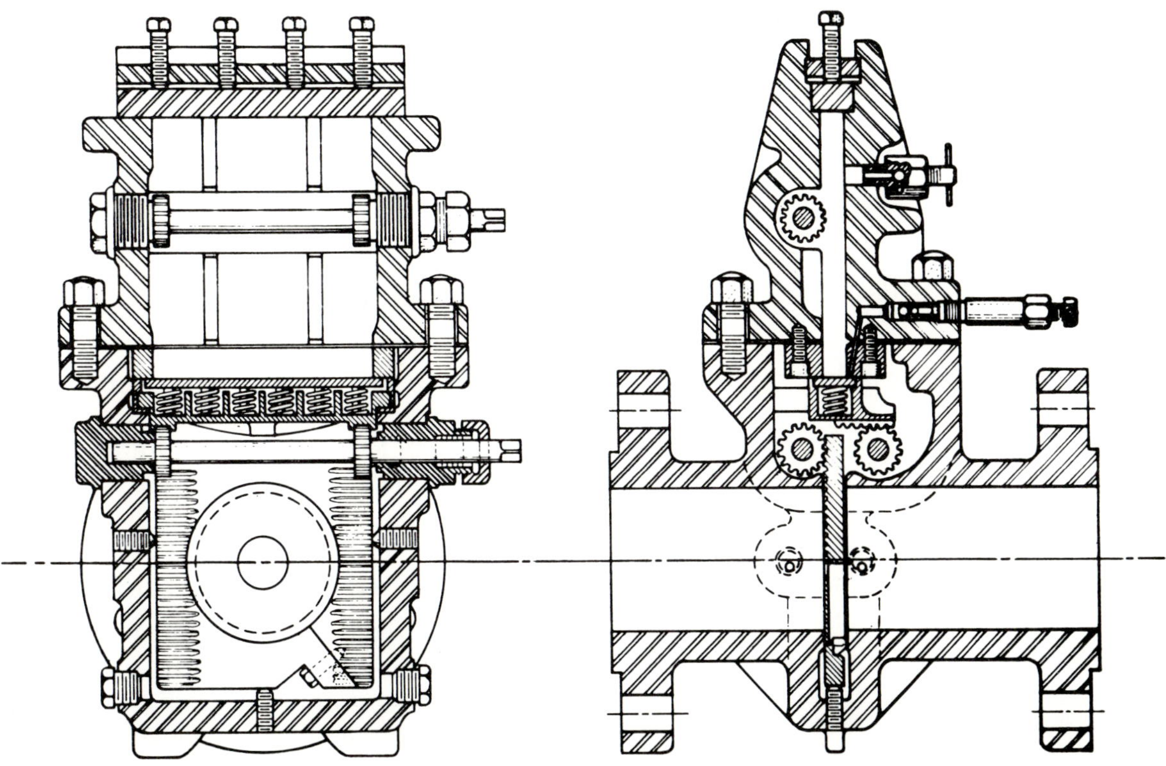

Figure 9.4. Bypass-type orifice flange fitting

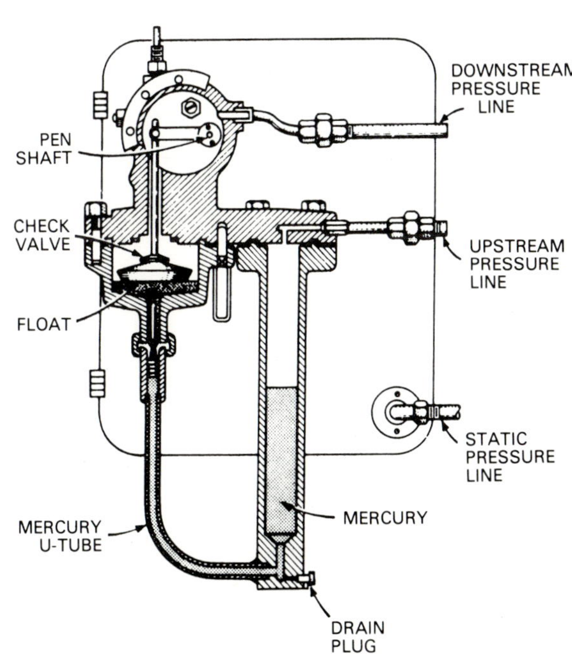

Figure 9.5. Differential-pressure measuring device using mercury manometer

Secondary element (recorder)

Orifice recorders normally include devices for measuring both differential and flowing pressures. Three types of differential-measuring devices are generally used. These are the mercury manometer (fig. 9.5), the bellows (fig. 9.6), and the force-balance

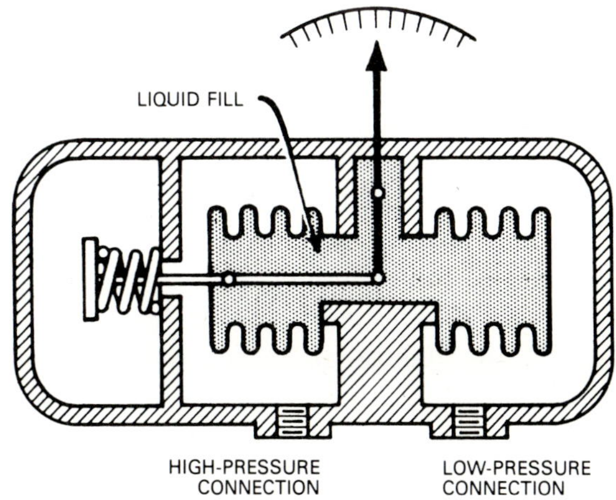

Figure 9.6. Schematic view of differential-pressure measuring device using bellows

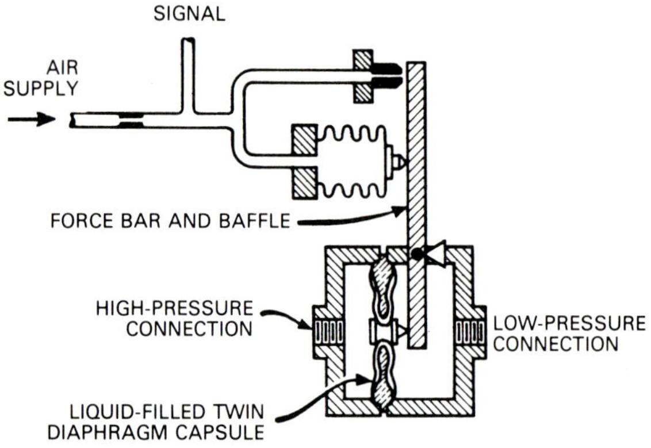

Figure 9.7. Schematic of differential-pressure measuring device using force-balance differential gauge

(fig. 9.7). Some flow or orifice calculations for measurement of gas will have one or more factors that will be affected by the type of meter used. AGA Report No. 3 lists such factors for mercury and bellows meters. Force-balance transmitters will use the factors listed for bellows meters. Flowing (or static) pressures are measured by conventional Bourdon tubes. In ordinary chart recorders, both differential and static pressure signals position separate ink pens that record the pressures on scaled circular chart paper. The charts are driven normally by 1-day or 8-day clock movements.

For gas service, a commonly used rule states that the differential range of the meter in inches of water should not exceed the static pressure in pounds per square inch absolute. That rule may require modification to conform to the characteristics of a particular application. Most gas measurements are made by use of differential-range meters of 0 to 10 in., 0 to 20 in., 0 to 50 in., or 0 to 100 in. Common practice is to select a static range that will permit recording in the upper half of the chart. For example, use 0- to 1,000-psi range for 700-psi or even 900-psi static pressure.

Temperature determination

Thermometer wells should be placed downstream of the orifice plate not nearer the plate than dimension *B* in figure 9.2. If it becomes necessary for the well to be placed upstream of the orifice plate, it should be upstream of dimension *A* or *A′*. If indicating thermometers are used, they should be protected against breakage, and the readings should be observed and posted on meter charts as frequently as may be practical but at least twice each chart period—that is, when the chart is placed and when it is removed. Recording thermometers should be used on most gas-sales meters. Common types are described in chapter 8. Recorder temperatures during periods of no flow should be disregarded in volume computations.

Specific gravity determination

The most commonly used method for the field determination of the specific gravity of a gas is by some form of weighing such as the Edwards balance. This instrument compares the density of the gas being tested with the density of air. The calculation of specific gravity is based on the principle that the densities of two gases at base pressure are in inverse ratio to the pressures that provide equal buoyant forces for the two gases. This method of specific gravity determination is used at locations involving both small and large volume sales and purchases of gas.

The recommended equipment for use in obtaining gas gravities by the balance method is as follows:

1. the Acme Senior gas gravity balance (fig. 9.8), for use by measurement specialists and at all locations where large volumes of gas are tested, and
2. the Acme Junior gas gravity balance, for use by measurement specialists working with small volumes of casinghead gas.

Two recognized technical standards are listed for reference. The first is recommended for use.

1. "Standard Method for Determining the Specific Gravity of Gases," Gas Processors Association Standard 3130

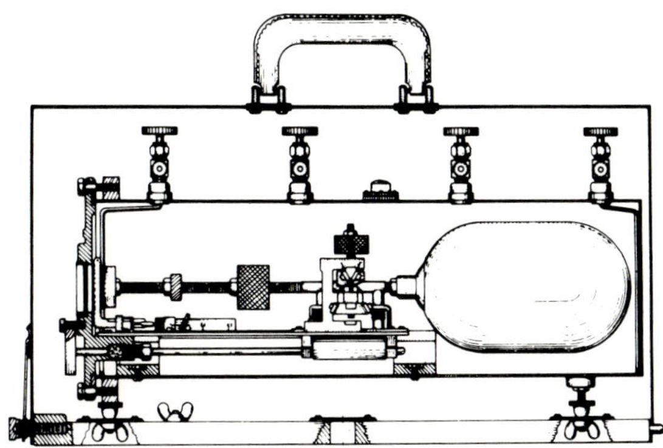

Figure 9.8. Acme Senior gas gravity balance

2. "Tentative Standard Methods for Determining the Specific Gravity of Gases," W.G.P. & O.R. Association's TS-391

The proportional torque differential indicator operates on the principle that the torque produced on the shaft of an impulse wheel activated by a jet of gas is proportional to the specific weight of the gas. The difference in torque produced by air and the tested gas is shown on an indicator scaled to read in terms of specific gravity. Recommended equipment is the Portable Ranarex indicator (fig. 9.9). This indicator is for determining specific gravities applicable to allocation meters.

Recording gravitometers may be used when a continuous record is desirable or required. The types most frequently used are those making use of weighing methods (fig. 9.10) and those using the momentum method.

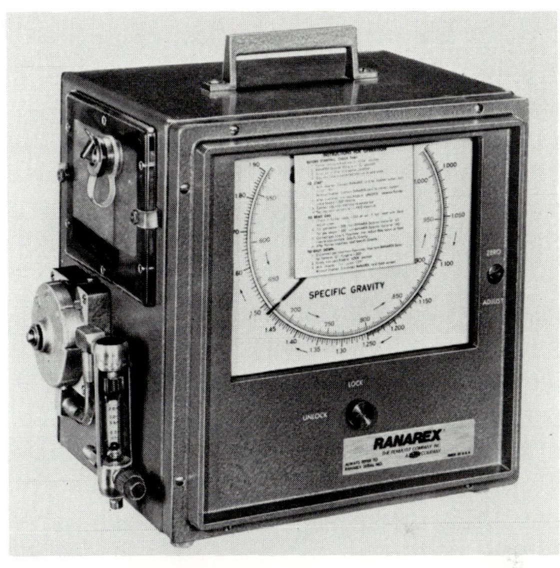

Figure 9.9. Portable gas gravitometer (Courtesy of Ranarex Instrument Company)

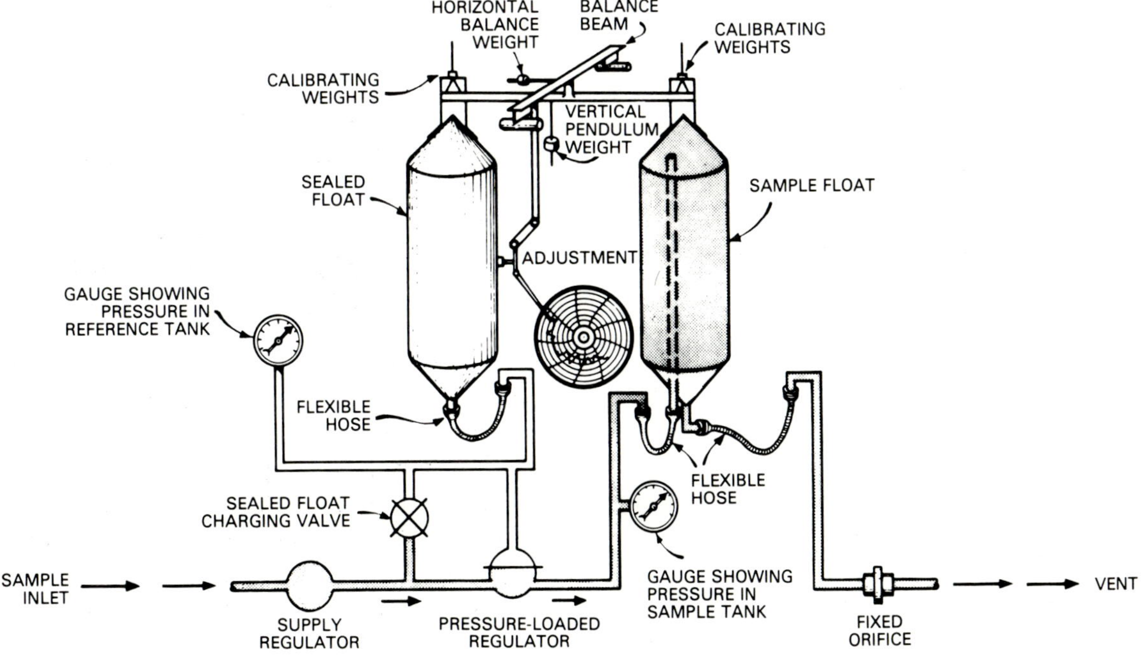

Figure 9.10. Schematic diagram of a recording gravitometer having gas-filled drums suspended from a balance beam

Meter piping

The differential lead lines that connect the instrument to the orifice flange or pipe may be ¼-inch nominal diameter pipe for lengths up to 15 feet and ½-inch nominal diameter pipe for lengths up to 40 feet. Tubing of equivalent internal diameter may be used in place of pipe. Distances greater than 40 feet are feasible but are not recommended. Where such distances are necessary, either pneumatic or electric transmission should be considered. Both the upstream and the downstream orifice connections should be made from the same side of the flange or fitting, preferably from the top (with a fitting placed on its side). In no case is it recommended that differential connections be made from the bottom of the fitting, flange, or line. Differential lines should be carried together, with a slope as great as possible but not less than 1 inch/foot. If the orifice is above the meter, the lines should slope up to the orifice. If the orifice is below the meter, the lines should slope down to the orifice.

Measurement calculations

Basic flow equation. In measuring the flow of gas with orifice meters, the flowing, or static, pressure of the gas as well as the differential pressure across the orifice must be obtained. Their relationship to rate of flow is expressed by the following equation:

$$Q_h = C'\sqrt{h_w p_f} = \text{rate of gas flow, cu ft/h at base conditions}$$

where

C' = orifice flow constant, corrected for operating and base conditions
h_w = differential pressure across orifice, in. of water
p_f = static pressure, psia

The product of the square roots of the differential pressure and static pressure absolute $\sqrt{h_w p_f}$ is commonly referred to as the *pressure extension.* Pressure extension tables are available in book form as a convenience in calculating gas flows.

Orifice flow constant. The orifice flow constant, C', is obtained by multiplying a basic orifice flow factor, F_b, by various correcting factors that are determined by the operating conditions, contract requirements, and physical nature of the installation. It is expressed in the following equation:

$$C' = (F_b)\,(F_{pb})\,(F_{tb})\,(F_g)\,(F_{tf})\,(F_r)\,(Y)\,(F_{pv})\,(F_m)\,(F_{sl})\,(F_e)$$

where

F_b = basic orifice flow factor, cu ft/h
F_{pb} = contract pressure base
F_{tb} = contract temperature base
F_g = specific gravity factor
F_{tf} = flowing temperature factor
F_r = Reynolds number (viscosity) factor
Y = expansion factor
Y_1, based on the upstream static pressure
Y_2, based on the downstream static pressure
Y_m, based on a mean of the upstream and downstream static pressures
F_{pv} = supercompressibility factor
F_m = manometer factor for mercury meter
F_{sl} = seal factor, applicable only to mercury meter sealed with a liquid
F_e = orifice plate expansion factor.

All orifice meter factors or coefficients should be selected from or calculated by use of AGA Report No. 3 unless the use of another method is required. A definite procedure for the calculation of coefficients can be established to ensure that all such coefficients will be calculated in the same manner and that, starting with the same basic data, any two persons using this procedure will arrive at identical answers for each coefficient.

Most natural gas streams contain water, and some of these streams are saturated with water. Errors in gas measurement and in gas composition may result if the presence of water vapor in the gas is not properly accounted for. The errors in volume measurement will be insignificant at pressures above 200 psig, but at lower pressures the errors can be substantial. The same is true of errors in composition as determined by analysis, although these errors are several times greater than the errors in volume.

Measurement error will not usually result because of water vapor in the gas if a field-determined specific gravity is used in the gas volume computation and if this gravity was determined on the wet sample. However, when a specific gravity is calculated from an analysis of the gas, the analysis is customarily reported on a dry basis and must be converted to a wet basis before a wet specific gravity can be obtained. Once the water content is determined, the volume of water obtained may be subtracted from the wet volume of gas to arrive at the gas volume expressed on a dry basis.

Accuracy of measurements. While it is impractical to list all the various factors that may affect the accuracy of measurement obtainable with differential-type flow instruments, it will perhaps be of some value to list the more common factors as sources of constant errors and variable errors. The classification is arbitrary because, under certain circumstances, the variable errors may result in constant or nearly constant errors, whereas several of the errors listed as constant may prove to be variable. The purpose of the lists is to provide a summary of possible sources of errors that may be helpful in locating and correcting errors and thereby obtaining better measurement. In most cases, corrections can easily be made mechanically or through adequate maintenance. Special situations such as freezing and pulsation will be discussed in more detail.

References may be found that indicate the relative effect or amount of error caused by many of these factors. However, it is not recommended that such data be used as a basis for applying corrections to flow calculations when these factors are present or suspected.

Flow measurement is often used as a basis for control only. As a general rule, the more accurate measuring installations will give more accurate control. However, in most cases, satisfactory control can be expected as long as the errors remain constant.

The constant errors include the following:

1. incorrect information as to the bore of the orifice plate;
2. contour of the orifice plate (convex or concave);
3. dullness of the orifice edge;
4. thickness of the orifice edge;
5. eccentricity of the orifice bore in relation to the pipe bore;
6. incorrect information as to the pipe bore;
7. excessive recess between the end of pipe and the face of orifice plate; and
8. excessive pipe roughness.

The variable errors include the following:

1. flow disturbances caused by insufficient length of meter tube or irregularities in the pipe, welding, and so forth;
2. incorrect locations of differential taps in relation to the orifice plate;
3. pulsating flow;
4. progressive buildup of solids, dirt, and sediment on the upstream side of the orifice plate;
5. improper check-valve operation;
6. accumulation of liquid in the bottom of a horizontal run;
7. liquids in the piping or meter body;
8. changes in operating conditions from those used in the coefficient calculations (i.e., specific gravity, atmospheric pressure, temperature);
9. incorrect zero adjustment of the meter;
10. nonuniform calibration characteristic of the meter;

11. corrosion or deposits in the meter range tube or float chamber;
12. emulsification of liquids with mercury;
13. dirty mercury;
14. incorrect arc for meter pens;
15. formation of hydrates in meter piping or meter body;
16. leakage around the orifice plate (applies to orifice fittings);
17. wrong range on chart;
18. incorrect time for rotation of chart;
19. excessive friction in the meter's stuffing box;
20. meter not level (mercury-type only);
21. excessive friction between pen and chart; and
22. overdampening of the meter response.

Processing meter charts. Charts should first pass through the hands of the operating location staff and, when necessary, a measurement specialist who will note on the back of the chart the gravity, temperature, supercompressibility factor, and any other information affecting the volume, and will verify the field data imprinted on the back of the chart. On the basis of past history or familiarity with conditions at the meter, the measurement specialist will estimate volumes when the meter was inoperative because of a stopped clock, pen not marking, freezing, overranging, mercury blown, and so forth.

When it becomes necessary to estimate volumes because of an inoperative meter, contract provisions governing such estimations should be followed. Many sales contracts include a provision to the effect that if for any reason meters are out of service so that the amount of gas cannot be determined from chart computations, the gas delivered during the period in which the meter was inoperative shall be determined by using the first of the following methods that is feasible.

1. By using check-meter readings.
2. By correcting the error with mathematical calculations if such error is ascertainable by calibration or test. Such errors include a wrong-sized orifice plate, a plate placed in backwards, a differential pen not zeroed, a static pen not calibrated, a mercury level that is high or low, and so forth.
3. By estimating the volume by comparison with deliveries during a period when the meter was operating properly.

Testing and maintenance

In making a routine meter test, the following items may be included. A portable deadweight tester or test pressure gauge of known accuracy may be used to calibrate a static pressure element. The check should be made at high, low, and medium chart range.

Meter lines should be inspected for leaks. Obstructions other than hydrates in the piping may sometimes be removed by use of a solvent. Specific instructions for testing the differential calibration are included in the operating instructions booklet furnished with new meters. These procedures provide for comparison of the recording to a water manometer as a standard when equal pressure is exerted on the recorder and on the manometer.

For most replacement purposes and usually with new meters, spring-driven standard clocks are used. However, battery- and electric-driven clocks are becoming more common. Normally, a spring-driven clock will run fast during the winter and slow during the summer because of the expansion and contraction of metal parts. At the approach of cold weather, clocks should be cleaned. Care should be taken that clock-mounting screws are not too tight, causing the clock to run slow. The clock shaft should be protected carefully during cleaning operations. A bent clock shaft will not rotate the chart concentrically.

Differential and static recording pens should be cleaned, inked, and the time-arc checked. The static pen should be adjusted to record behind (i.e., lag) the differential pen approximately 15 minutes on 24-hour charts and 2 hours on 7-day charts. This adjustment is to make recordings compatible with chart-integrating devices and facilitate integration of the charts.

Mercury-type meters. When it becomes necessary to replace a worn differential shaft, the whole stuffing box and shaft should be replaced with a Teflon-bearing stuffing box. The allowable stricture of the differential shaft should be not more than 0.4 inch of water for high-pressure meters and 0.1 inch for low-pressure meters. If this frictional tolerance is exceeded, the stuffing box should be removed and the shaft lapped to the bearing with jeweler's rouge. Care must be taken that all rouge is removed, by several washings in solvent, after the lapping is complete. The stuffing-box lubricant should vary with the type of service and climatic conditions. Overpacking with grease may cause the shaft to bind.

The mercury must be clean. The mercury chamber should first be cleaned with an approved type of solvent. The mercury may be cleaned by straining through a chamois or coarse cloth or washing in a 10 to 20 percent solution of nitric acid and rinsing with clear water.

The float must be properly centered and should travel freely. Erratic float travel may be due to a worn pin or to the manometer not being level.

The check valve (inside meter bodies) must operate freely.

It is extremely important that the manometer be level. If the instrument level is broken, a 6-inch machinist's level should be used across the top of one of the mercury chambers that has had the cover removed.

Bellows-type meters. Evidence of leakage of meter internal fill liquid should be checked for. When leaks occur, it is not recommended that the repairs and refilling be attempted in the field but that the faulty bellows unit be exchanged for a new or factory-overhauled unit.

Liquid and foreign material accumulations in the bellows housing should be checked for. The bellows should be overranged and underranged, and sticking overrange seal valves noted.

Evidence of internal and external corrosion should be investigated.

Excessive meter dampening should be noted.

Orifice plate. Orifice plates should be inspected to ascertain that—

1. the upstream edge is sharp and without a wire edge;
2. the face of the plate is flat;
3. the face of the plate is smooth and without pits;
4. the correct size of the bore, measured by micrometer, is stamped on the plate; and
5. no dirt or ice has collected against the orifice plate.

Orifice fittings. If the meter tube is equipped with an orifice fitting, the following observations and operations should be made.

1. Packed glands must be kept tight.
2. The moving parts must be lubricated.
3. The plate carrier should be removed for inspection on a routine schedule.
4. Moving parts should be actuated to prevent them from becoming frozen.
5. Checks should be made to see that the orifice plate holder is not frozen into place or that water does not collect about the isolating valve and burst the upper chamber during very cold weather.

Measurement problems

Freezing. Hydrate formations at the orifice, in meter piping, or in the meter chamber may occur when the temperature of the wet gas being measured falls below the hydrate temperature. The chart, on which recordings have been made while the meter was partially frozen, should have a full explanation written on the face of the chart and estimated static and differential pressures should be lines drawn in.

Preventive measures include—

1. elimination of piping leaks;
2. installation of line heaters;
3. installation of a heated meter house;
4. dehydration of gas;
5. use of inhibitors;
6. enlargement of the meter piping and valves to ½ inch maximum; and
7. replacement of needle valves with plug or gate valves.

Pulsating flow. Pulsations in a pipeline originating from a reciprocating system or some other similar source consist of sudden changes in both velocity and pressure of the flowing fluid. The pressure changes are the more apparent and resemble low-frequency sound waves traveling in the flowing medium with a velocity independent of the velocity of the flowing fluid. The most common sources of pulsation involved in gas measurement are—

1. reciprocating compressors, engines, or impeller-type boosters;
2. pumping or improperly sized pressure regulators and loose or worn valves;
3. irregular movement of quantities of water or oil condensate in the line; and
4. intermitters on wells and automatic drips.

Reliable measurements of a gas flow with an orifice meter cannot be obtained when appreciable pulsations from any source are present at the point of measurement. No way has been found to determine or predict correction factors to compensate for such errors.

In order to obtain reliable measurements, it is necessary to suppress the pulsations. In general, the following methods are valuable in diminishing pulsation and its effect on orifice flow measurement:

1. locating the meter tube in a more favorable location with regard to the source of pulsation, such as at the inlet side of regulators, or increasing the distance from the source of pulsation;
2. inserting capacity (volume), restriction, or specially designed filters in the line between the source of pulsation and the meter tube in order to reduce the amplitude of the pulsation;
3. operating at differentials as high as is practicable by replacing the orifice plate in use with an orifice plate having a smaller orifice or by concentrating flow in a multiple-tube installation through a limited number of tubes;
4. using smaller-sized tubes and keeping essentially the same size of orifice while still maintaining the highest practical limit on the differential.

As yet no instrument has been developed that will give precise quantitative effects of pulsation on flow measurement. However, instruments have been developed, both mechanical and electrical in nature, that indicate the presence of pulsation that could affect the measurement.

Slugging. The conditions commonly called slugging refer to a liquid (water, oil, or condensate) accumulation in a gas line. In low-pressure lines, the liquid will gather at a low place in the line and restrict the passage of gas until enough gas pressure has accumulated to blow through the liquid. In a high-pressure system, the liquid will sweep up to and through the orifice. Both conditions produce erratic recordings and inaccurate measurement of an undeterminable extent. The use of drips or liquid accumulators of various types is often the only feasible solution to the problem. When chokes are used on wet-gas lines, they should be placed downstream of the orifice.

Sour gas. Freezing and corrosion are two frequent problems in measuring gas containing hydrogen sulfide. Corrosion in a closed line free of air and water is negligible, and most meter corrosion is due to hydrogen sulfide in the surrounding atmosphere. Possible remedies include the following changes in equipment:

1. Static spring: 316 stainless steel is generally satisfactory.
2. Differential pen shaft: Teflon bearings that are unaffected by hydrogen sulfide are used. Lubrication with a silicone lubricant is helpful.
3. Pen: Self-feeding pens give better service, as they are more closely sealed against the atmosphere.
4. Clocks: Vaporproof clocks are essential. The rubber seal should be coated with varnish. The winding stem and chart hub stem should be coated with grease.
5. Seal pots: Seal pots are essential for protection of mercury in mercury-type meters. Some recommended sealing fluids are ethylene glycol or glycol-base antifreeze compound with 40 percent water. An inhibitor of 4 ml of 25 percent formaldehyde per gal may be added. The orifice factor must be properly corrected when sealing fluids are used.

Other methods of gas measurement

Displacement meters. Displacement meters are often referred to as positive displacement meters, since they afford a positive volume in cubic feet at flowing conditions regardless of the flowing temperature or the specific gravity of the gas. Displacement meters used in the field are generally of two types: the rotary or impeller type and the slide-valve diaphragm type. Turbine meters are sometimes included in this category.

The rotary and diaphragm types of displacement meters contain measuring elements of known volume and valve arrangements to channel the gas into and out of the measuring elements. They are also equipped with a counter or index dial to count the number of times the volume container or element may have been filled during the measurement process.

Since displacement meters measure the volumetric displacement of the gas at flowing conditions, it usually becomes necessary to record the pressure or provide some other means of adjusting the meter displacement to the base pressure of the defined measurement unit. Likewise, when required, flowing temperature adjustments to the base temperature must be made.

Depending on the class of service, several types of volume recording or totalizing register devices may be installed on displacement meters. These include the following devices.

1. The dial index is used when constant pressure can be maintained on a meter.
2. The dial index and recording pressure gauge is used when constant pressure cannot be maintained on a meter. Chart pens record the gauge pressure on the calibrated portion of the chart and the volume on the outer edge of the chart. The latter registration is by means of a series of waves, or loops; each complete wave or loop indicates that one (or ten) Mcf has passed through the meter at the pressure indicated by the pressure pen. Volumes computed by counting the loops will check very closely with the dial index reading.
3. The base pressure corrector is used to correct the volume of gas passing through the meter to the base pressure. This volume is registered on an accumulating counter. No pressure correction is needed for calculating the volume recorded.

Volumes as recorded by positive meters are independent of specific gravity correction. For a dial index and recording pressure gauge, the recorded volume must be corrected for pressure and, if necessary, for temperature. The base pressure corrector needs only the temperature correction.

It is recommended that field personnel not attempt field repairs on displacement meters, although repairing and replacements may be made in the field by trained personnel with proper tools. It will be sufficient that measurement personnel check to see that the meter is operating within its rated capacity, since increased speed means increased friction and wear. Of the several standard methods of testing displacement meters in the field, it is recommended that a critical flow prover be used when the gas pressure can be maintained above 15 psig.

Mass-flow meter. The orifice meter may be used to measure gas on a mass-flow basis in pound units. The technique requires determination of the density of the gas being measured and the differential pressure across the orifice. Equipment is now available to measure the density of a flowing gas. The equipment is called a densitometer and is available for both electric and pneumatic applications. In the use of the orifice meter for making mass measurements, the densitometer is substituted for the static-pressure element and makes determination of the specific gravity and supercompressibility corrections unnecessary.

The following equation from AGA Report No. 3 may be used for mass flow computations:

$$W = 1.0618 F_b F_r Y \sqrt{h_w \gamma}$$

where

W = flow rate, lb/h

F_b = basic orifice factor

F_r = Reynolds number factor

Y = expansion factor

h_w = differential pressure

γ = specific weight of gas at flowing conditions, lb/cu ft.

Another mass meter is a volumetric device called the vortex-velocity mass-flow meter, where weight flow is determined from the volume flow in cubic feet at flowing conditions and the specific weight in pounds per cubic foot.

Still a third mass-flow meter makes use of a specially designed gyroscope that is caused to precess around its major axis when the torque from the turbine is applied to its minor axis. The speed of the precession is directly proportional to the weight-flow rate. The cumulative mass flow is recorded on a counter operated by the precession.

Turbine meter. Several makes and designs of turbine meters are currently available. These make use of the flowing gas as a force imparted to a bladed rotor. By use of appropriate gearing, revolutions of the rotor may be converted to volume. Accuracy curves are usually developed for each turbine meter, and proving or calibration techniques are being developed. Filters ahead of turbine meters are almost a necessity to permit sustained accuracy and trouble-free operation. Progress in this type of measurement indicates that many turbine meters may soon be in use.

Elbow meter. Centrifugal force in the curve of a pipe elbow can be used to measure flow. For accuracy, the elbow should be calibrated by using some other acceptable measurement as a standard. Accuracy is not usually the objective when elbow meters are used. Relatively little pressure loss or differential pressure is created. As a result, the meters are used primarily for control or other operation purposes.

Natural gas liquids measurement

Orifice meter

Field measurements of natural gas liquids are accomplished by the conventional gauging of tanks and by use of various metering techniques. The orifice meter is sometimes used. Installation and operation requirements are about the same as for gas. Some of the basic data are used in the calculation of orifice meter recordings into pound or gallon units that are used for gas. However, there are different factors to be applied.

For measurement into gallons, for example, the American Meter Company's *Orifice Meter Constants: Handbook E-2* may be used. The equation for this handbook is as follows:

$$Q_h = C'\sqrt{h_w}$$

where

Q_h = rate of liquid flow, gal/h

C' = orifice constant ($F_b \times F_{gt} \times F_{sl} \times F_r$)

h_w = differential pressure, in. of water

For determination into pound units the following equation from *Principles and Practices of Flow Meter Engineering,* 8th edition, published by Foxboro Company in 1961, may be used:

$$W = S\,N\,D^2\,F_a\,F_m\,F_c\,F_p\,\sqrt{G_f h_w}$$

where

W = rate of flow given in pounds per 24 hours

S = a value determined for the bore of the orifice and the internal diameter of the metering tube

N = combined constant for weight-flow measurement equal to 68,045 when W is in pounds per day

D = inside diameter of the meter tube given in inches

F_a = correction for thermal expansion of the primary device (orifice) and assumed to be 1.000 as long as the temperature of the mixture at measurement remains between 23°F and 99°F, using a steel orifice plate

F_m = manometer factor that for a bellows-type meter is 1.000

F_c = viscosity factor, usually assumed to be 1.000

F_p = correction for compressibility of the liquid. This factor is sometimes combined in the definition of G_f

G_f = specific gravity of liquid stream at a flowing temperature and pressure as determined by gravitometer readings

h_w = differential pressure in inches of water as recorded by a bellows-type orifice meter, using flange taps with flowing liquid in lead lines and bellows chambers.

This equation, for simplicity, may be converted to:

$$W = 68{,}045\ S\,D^2\,\sqrt{G_f h_w}$$

In each of these equations, flow coefficients corrected pursuant to contract or specified conditions and units must be developed. Each coefficient is then multiplied by the square root of the chart differential pressure in inches of water to get the flow rate per hour. Such a rate times the number of hours gives the quantity measured during a given period of time.

Using the orifice meter, as above, or any of the following briefly described liquid measurement methods, it is essential that the fluid must be in a single phase, that is, all liquid at the point of measurement.

Positive displacement meter

Liquid measurement by use of positive displacement methods can be very accurate when appropriate corrections are applied. Examples of procedures to be followed are included in API Standard 1101, *Measurement of Petroleum Liquid Hydrocarbons by Positive Displacement Meter.* Other publications containing information relating to liquid measurements include API Standard 2502, *Lease Automatic Custody Transfer.*

In API and other manuals, it is noted that the equipment requires periodic maintenance and regular proving. The frequency of these operations is dependent on the quantity of fluids being metered. Sampling is critical. Continuous samplers are recommended and should include agitation as a means to enable accurate impurity determination.

Sizing of meters is also critical, as are other considerations listed in the API Standards.

Turbine meter

The turbine meter, described briefly in the coverage of gas measurement above, is very good for liquid measurement if there is no emulsification of the fluid at the point of measurement. The manufacturers furnish calibration and operation data with each unit, as well as calculation procedures for gallon or pound units.

Two-phase flow

In the above coverage of gas and liquid measurements, emphasis has been given to the fact that accuracy suffers if the fluid is not in single phase. However, it is frequently necessary to make measurements for operation and allocation purposes when the fluid is two-phase, that is, both gas and liquid.

If measurement is required of a two-phase stream, certain precautions may be taken to arrive at reasonably acceptable measurements with orifice meters.

1. Pressure and temperature must be kept as high as feasible at the meter.
2. A free-water knockout should be used ahead of the meter.
3. A vertical meter run may sometimes improve the differential-pressure relationship to the volume.
4. Test data from periodic full-scale separator tests are used to determine coefficient or meter factor.
5. Manifold lead lines to bottom of bellows-type meter should be connected with self-draining pots installed above orifice fitting.

Natural gas testing

Gas testing is an important phase of the field handling of natural gas. Although not normally considered as such, it actually is a form of measurement—the determination of the liquid hydrocarbon content of the gas. The results of such tests may determine how much a seller receives for his gas or how much of the liquid is allocated to several leases that may be connected to a common system. Royalty payments may also be directly affected by the results of such tests.

Although complete testing is not always done in the field, sampling is. The results of the analysis can be representative of the gas stream being tested only to the extent that the sample is so representative. Thus, sampling procedures must assure that containers and testing equipment are purged of all extraneous gases or vapors. The gas-sampling point must be located so that liquids, if condensation takes place, cannot be drawn directly into the sample container. Special equipment must be provided in the event that a two-phase stream is being sampled. Leaks in any part of the sampling or testing equipment cannot be tolerated. Care must also be taken to prevent condensation in the lines connecting a sample container to the source of the gas being sampled.

A number of methods are used to test gas, depending upon its composition, the gas-sales contract provisions, and the use to which the results are to be put.

Charcoal tests

The charcoal test is largely limited to gas which is low in C_4+ content—usually 1.0 gallon per thousand cubic feet (GPM) or less. The gas is drawn from the stream being tested and passed over activated charcoal. Hydrocarbons are selectively adsorbed to the charcoal, with the heavier components replacing the lighter. After the adsorption process is complete, the charcoal is taken to a laboratory, where the adsorbed liquid is driven off by heat, condensed, and measured. The results are converted to GPM of gas. The results obtained from this testing method can be affected substantially by such things as the rate and length of time the gas sample is passed over the charcoal, the quality of the charcoal, and the techniques used in recovering the hydrocarbons from the charcoal. It is used where other methods of testing either give unsatisfactory results or are too expensive.

Compression testing

Compression testing is used extensively on casinghead gas that has a C_4+ content of above 0.5 – 1.0 GPM. The gas sample is compressed to 250-psig pressure in a portable compressor and then cooled in an ice-water bath or in a refrigerated condenser to near 32°F. The liquid so condensed is converted to GPM in the same manner as in the charcoal test procedure. Compression testing is quite reliable. It is essential to have all equipment in top operating condition if repeatable test data are to be obtained. This is the one form of gas testing that is done completely at the field location.

Fractional analysis

Fractional analysis is used in those cases where a knowledge of gas composition is needed. It is also used in testing two-phase gas-condensate streams. In the procedure, a sample of the gas stream is obtained in a metal container and shipped to a laboratory where a fractional analysis is performed. In the case of two-phase streams, it is necessary to separate the gas and liquid in a separator and obtain two samples—one gas and one liquid. The rate of production of each phase is obtained at the time of sampling so that a composite sample can be calculated after each sample has been analyzed.

It is obvious that extreme care must be used in obtaining samples to prevent any contamination. Also, in the case of two-phase streams, precise data on production rates, pressures, temperatures, and so forth are essential.

Laboratory data on fractional analysis will generally show not only the composition in percent of each hydrocarbon present through hexanes or heptanes but will show also the GPM by component (usually beginning with ethane) and the heating value of the gas.

Field procedures for the first two testing methods require the use of a test car. Usually representatives of the buyer and seller meet at the test location and check each other, verifying the finally accepted result. The frequency and time of testing is set out in the contract covering the sale. When analyzing is done in a laboratory, as in the third testing method, an independent commercial laboratory is usually employed. These firms perform the entire test—sampling as well as analysis.

For more detailed information, refer to the following publications issued by the Gas Processors Association.

AGA-GPA Code 101-43: Standard Compression and Charcoal Tests for Determining the Natural Gasoline Content of Natural Gas

GPA 2166-68: Methods for Obtaining Natural Gas Samples for Analysis by Gas Chromatography

GPA 2165-75: Method for Analysis of Natural Gas Liquid Mixture by Gas Chromatography

GPA 2161-72: Method of Analysis for Natural Gas and Similar Gaseous Mixtures by Gas Chromatography

GPA 2265-68: Method of Determination of Hydrogen Sulfide and Mercaptan Sulfur in Natural Gas

Appendix A
Gas facility maintenance

Generally speaking, the two main types of maintenance programs are (1) unplanned and (2) preventive. In the first type, repairs are not made until there is a breakdown or some sort of unexpected shutdown. In the second type, equipment checks are made on a planned basis with the objective of making needed repairs before a failure or shutdown occurs. Preventive maintenance programs can be quite simple or very elaborate in accordance with the operator's desires. Such planned programs are based on actual performance history, and the keeping of repair and operating records is essential. These records may be relatively simple or very detailed, depending upon the elaborateness of the program in effect. As an example, some maintenance programs are designed for major equipment items (i.e., large compressors or pumps), and preventive maintenance work is done at arbitrarily determined intervals, such as once a year or once a month or once a number of operating hours. In such cases, only rather general records of performance are required. At the other extreme, some programs utilize computers that compile and process the operating and repair data needed to provide the basis for the program. In such cases, the records prepared by operating and maintenance personnel must be very accurate and as detailed as the plan requires. Any number of plans fall in between these two extremes.

A decision about the type of maintenance program to be used in any particular operation has to be made by the operator or company involved after some sort of economic evaluation has been completed. In this evaluation, the operator considers production losses due to downtime, operating expense, and maintenance costs (including wages, repair material, cost of special tools, etc.) for several types of maintenance programs to determine which one has the least total cost. Preventive maintenance programs of some sort are selected by most operators because they have determined that repair costs under this type of program are no greater than those for the unplanned type (and are often less in the long run), while at the same time downtime with its attendant production losses is greatly reduced. In many cases, however, it is more economical to let equipment run until it fails and then replace it with repaired or new material that is kept available. Small pumps, magnetos, and spark plugs are examples of such equipment, which is either readily available at stores or economically stored on the job site. In reaching his decision in such cases, the operator must take into account the investment cost of having spare equipment on hand, production losses, and the maintenance program itself. Very sophisticated electronic devices are available to detect potential failures well before they happen. Their use reduces the need to shut equipment down for visual inspection. The operator must consider the cost of the equipment and its operating and maintenance cost in determining whether to use these devices or to employ a less elaborate program.

Preventive maintenance is an absolute must for safety and protective shutdown devices and alarms. In order to be assured of reliable operation, this type of equipment must be checked on a very frequent schedule. Unless the operator has a reasonable assurance that it will work when an emergency condition occurs, the money spent for its installation will have been wasted. The lives of operating personnel and the general public may be jeopardized if these devices are not in good operating condition at all times, and the equipment ostensibly being protected is also endangered.

Some of the factors to be considered in making the economic evaluation necessary to the proper selection of a maintenance program are as follows.

Load. Obviously, heavily loaded equipment requires more attention than that which is lightly loaded. A preventive maintenance program will definitely keep downtime at a minimum and improve production. Electronic testing devices are particularly helpful in such cases, since shutdowns for maintenance are made only when potential trouble is indicated by the testing equipment.

Availability of repair or replacement parts and equipment. At times it takes several weeks to obtain important repair parts. If the investment cost is reasonable, repair parts can be stocked at the site of the operation. Otherwise, frequent checking regarding wear, fatigue, and so forth needs to be made to establish ahead of time when a replacement or repair should be made. Such checking provides the basis for making the repair on a planned basis and at the most opportune time. This should reduce investments in repair parts, minimize production losses, and perhaps eliminate excessive repair costs that often occur when equipment completely fails.

Standby equipment. As a facility gets older, it often becomes somewhat underloaded. The net result may be that the equipment idled by the reduced load in effect becomes standby equipment. Any maintenance program should take this condition into account. A much less rigid maintenance schedule can be utilized, since production losses should not be such a dominant factor. The installation of standby equipment in new facilities can be justified only if the installation cost can be amortized quickly by savings in maintenance cost plus reduction in production losses.

Life of operation. Some equipment has little or no salvage value once it has been installed. Buildings and buried pipelines are examples of such equipment. If the life of the equipment is approximately the same as that of the project itself, there is little or no justification for maintaining it other than for appearance. Adequate testing procedures (e.g., testing for rate of corrosion in pipelines) are essential, and the life of the project must be established beyond question. Any established program should be reviewed periodically to determine whether it should be reduced in scope or discontinued altogether, as dictated by the remaining life of the project.

Operating conditions. Weather, isolation, and inaccessibility all must be considered in selecting a maintenance program. Moving machinery in dusty areas must have special attention. Rust and corrosion must be controlled in humid areas. Unattended equipment needs to have protective shutdown devices in good operating condition. The spare part supply becomes important when accessibility is a factor.

Environment. Prevention of all types of pollution is becoming more and more important. High operating efficiency as a rule reduces pollutants from equipment such as internal-combustion engines, and it can be obtained through good maintenance. Devices installed for the express purpose of controlling pollution need constant attention for proper functioning. Even an occasional spill is no longer condoned by regulatory bodies. The control of noise is becoming necessary, so the maintenance of the dampening devices is a continuous matter. A backfiring engine quickly brings complaints from nearby residents. Maintenance of this type of equipment is not justified economically; it is becoming one of the costs of doing business.

Hazards. Natural gas is safe enough when properly handled. Improperly maintained gas-handling equipment can be a threat to the public, employees, and the owner's property. Gas-handling equipment is one of the most important factors to be considered in choosing a maintenance program. Again, economics do not apply directly, for it is impossible to place a monetary value on a person's life or well-being. The equipment must be safely constructed, operated, and maintained.

Operations safety inspections. The following checks should be made daily:

1. Check for hydrocarbon leaks in piping and equipment. Pay close attention to leaks from vents and safety relief valves.
2. Check direct-fired heaters for hot tubes, flame impingement, and improper fuel-air mixing at the burner.
3. Check gas engines and compressors for excessive movement on the foundation and cylinder attachment to the frame.

Appendix B
Notes on gas processing plants

The facilities normally operated in the handling of gas in the field are those required to condition the gas only to make it marketable—removal of impurities, water, and excess hydrocarbon liquids, and control of delivery pressure through the use of pressure-reducing regulators or compressors.

Gas processing plants are usually designed to remove certain valuable products over and above those needed to make the gas marketable, that is, natural gasoline, butane, propane, ethane, and even methane in some instances. In order to do so, plant processes include many of the functions ordinarily performed by gas conditioning equipment, such as dehydration and H_2S removal. Thus, a plant may be considered as another gas conditioning facility. The equipment used is essentially the same as lease equipment for liquid removal, dehydration, and acid-gas removal. Plants usually provide fractionating equipment to separate the liquid hydrocarbons recovered into pure products or predetermined mixtures. When H_2S is removed from gas, a plant may include facilities to recover elemental sulfur from this impurity.

The decision as to whether or not a plant should be installed is often greatly influenced by the amount of gas conditioning needed. Since condensible hydrocarbon liquids cannot be tolerated by a gas transmission system, it is mandatory that these products be removed from the gas before it enters the pipeline. A plant can more efficiently remove these liquids and can recover a greater quantity of them than is possible with a smaller lease processing facility. The increased liquid production may well justify the entire gas processing plant installation economically, resulting in a substantially greater money return than would result from simple gas conditioning with the usual field facilities.

Plants are basic to the cycling of a retrograde reservoir, since the purpose of the operation is to recover the maximum quantity of liquids while maintaining reservoir pressure. Although the use of simple separators and compressors will recover most of the heavier condensate, recoveries can be greatly increased with a processing plant, and separation of the products into various components is possible.

Plants may be justified simply for the recovery of certain hydrocarbon components even if those components are only a very small part of the gas stream. As an example, a plant may be installed to process the gas of a transmission system for the recovery of a very small percentage of propane or ethane. The gas volume throughput of a transmission line is usually large, so the actual gallons of liquid propane or ethane recovered can be substantial. In such cases, large volumes are essential for the project to be economical. The product recovered often has a special value as chemical feedstock. Such plants are not installed to provide conditioning for the natural gas itself.

Plant size is governed by many factors. Quite small plants—20 to 50 MMcf/d—can be profitable in processing the fairly rich gas from an oil field or the gas from a rich gas condensate field. Plants of this size are often semiportable and may operate on a largely unattended basis. As noted earlier, where recoveries are small, large volumes of gas must normally be processed. Cycling plants may handle several hundred million cubic feet per day. The processing for sale of gas from several fields is often performed in a single centrally located plant, which again may be very large—perhaps having a throughput of several hundred million cubic feet per day.

Equipment found in gas plant operations is similar to that found in lease operation, the primary difference being the size and, perhaps, the mechanical design of the units. Typical types of equipment found in plants are absorbers, strippers, fractionators, heat exchangers, and aerial coolers or cooling towers.

Plant products

Plant products are normally considered to be those materials recovered from the natural gas stream that have value in addition to the residue gas itself. The type of operation determines the products to be saved. Cycling plants may produce everything from ethane to naphtha or heavy condensate, including motor fuels. Plants processing casinghead gas can produce ethane, propane, butanes, pentanes, and natural gasoline. Plants operating on very dry gas may produce only propane, butane, and natural gasoline. Sulfur, of course, may be produced by plants handling gas containing H_2S. The amount and kinds of products produced in a given plant will depend on composition of the feed gas and the contractual specifications of the residue gas delivered from the plant. Thus, ethane and propane production may be limited in order to maintain the Btu content of the residue gas.

In some new plants, the residue gas is liquefied for transportation when pipelines are not available or possible. This liquefied natural gas (LNG) is the product of a refrigeration process that requires operations at very low—cryogenic—temperature ranges (about −260°F).

Product specifications and methods of sampling and testing are published by the Gas Processors Association and the American Society for Testing of Materials. Many of the specifications can be found in the Gas Processors Suppliers Association *Engineering Data Book*. Sales contracts for both liquids and gas usually set out the specifications that must be met.

Hydrocarbon recovery processes

The recovery of liquid hydrocarbons from natural gas in a plant is accomplished by changing conditions of the gas so that the equilibrium between the various components is upset, causing some components to condense and others to vaporize in attempting to reach a new equilibrium. Pressure or temperature may be changed, a different material may be introduced into the gas stream, or a combination of all three changes may be established.

The early method of recovering liquid hydrocarbons from natural gas was compression and cooling. Engineers found that by compressing natural gas to higher pressures and cooling it to near-ambient temperature, certain hydrocarbon liquids were formed and could be separated from the gas stream. It was noted that this recovery could be predicted, using equilibrium vaporization constants and natural gas analysis. The compression and cooling process was and is by far the simplest method. However, it is not as efficient as some of the methods developed later. The compression and cooling method was normally limited to cooling by ambient air or by cooling water. A logical development from the conventional compression and cooling method was the use of refrigeration to reduce the temperature of the gas stream further and recover more of the product. Mechanical refrigeration systems with ammonia or propane as the refrigerant were the first types of refrigeration used. Of course, the early attempts met several problems associated with hydrate formation. Freezing occurred in the gas chiller and in the separator downstream of the chiller. Injection of methanol or a glycol solution into the gas stream greatly helped this problem. In some cases, wax and paraffin formation in the chiller caused problems that had to be resolved. In normal plant work, the low-temperature separators usually operate in the temperature range of −20°F to +20°F. Refrigeration systems include the use of freon, ammonia absorption, or, more recently, a turboexpander where a pressure drop is available in the natural gas stream or can be provided economically. Each system is used for the same purpose—to cool the gas stream to a temperature that will cause liquefiable hydrocarbons to condense.

The absorption system of hydrocarbon recovery (fig. B.1) employs the principle of introducing a different material into contact with the gas stream. Absorption oil, having a particular boiling range and molecular weight, is flowed countercurrent to the wet gas in a tray-filled absorber tower. The heavier components of the feed gas condense and flow to the bottom of the tower with the absorption oil. Dry residue gas flows out of the top of the tower. The rich absorption oil is heated in a still (i.e., a fractionating column) where the liquid fractions recovered from the gas are distilled from the absorption oil. The lean oil is returned to the absorber. The recovered liquids are then fractionated into the desired products—propane, butane, natural gasoline, and so forth. By applying refrigeration to both absorption oil and feed gas, recoveries can be improved. Recoveries of 40 to 60 percent of the ethane and 100 percent of the propane and heavier products are obtained with operating temperatures of −20°F.

The most recent development in low-temperature, high-recovery plants is the cryogenic plant that uses the turboexpander.

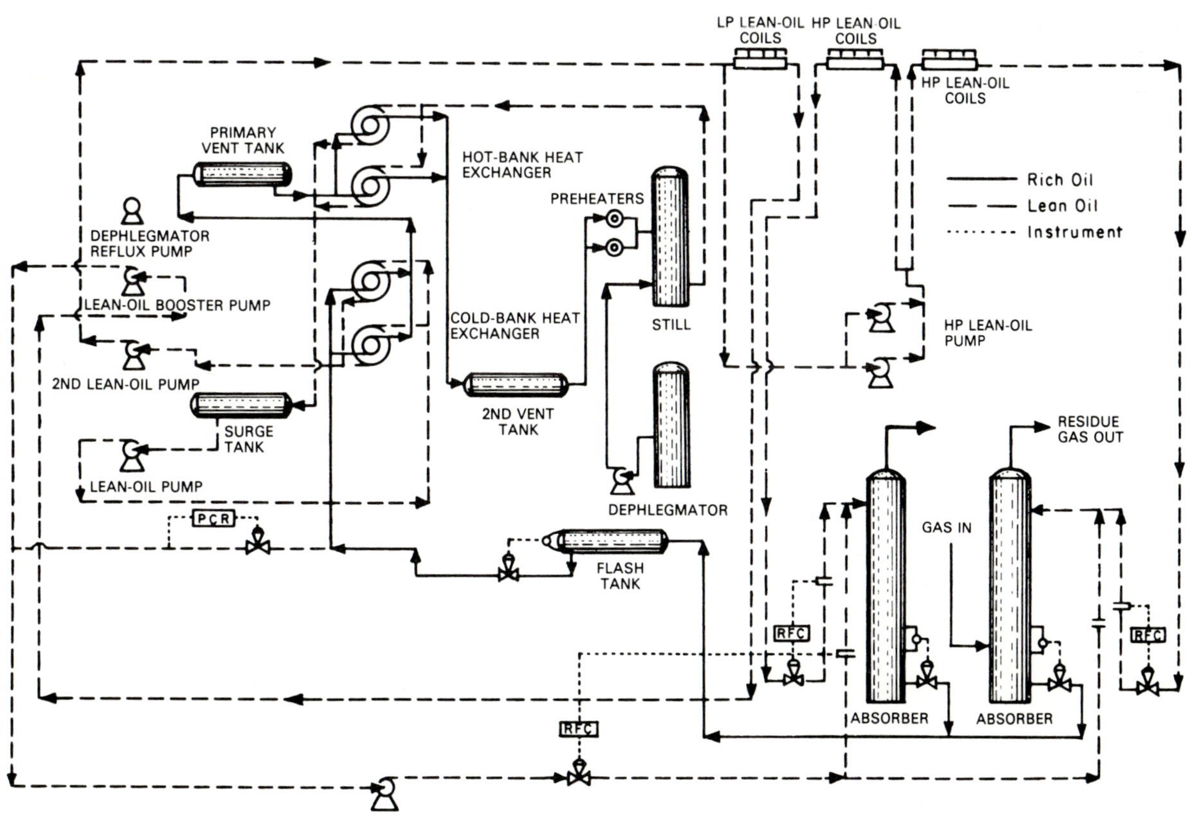

Figure B.1. Flow diagram of an absorption plant

In this type of plant, the gas is expanded through a turbine compressor from which it exhausts at extremely low temperatures in the range of −160° to −180°F. At these low temperatures, most of the gas except methane is condensed. The liquids are then fractionated to recover desired products in an ordinary fractionating system. Dehydration is important in any system employing low temperatures. In cryogenic plants nearly 100 percent dehydration is absolutely essential at all times.

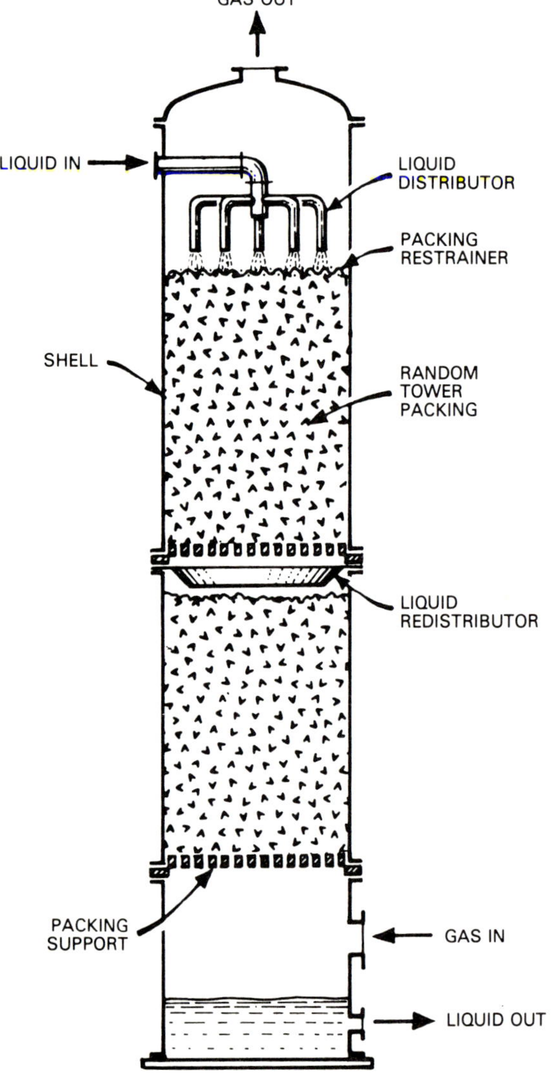

Figure B.2. Packed tower

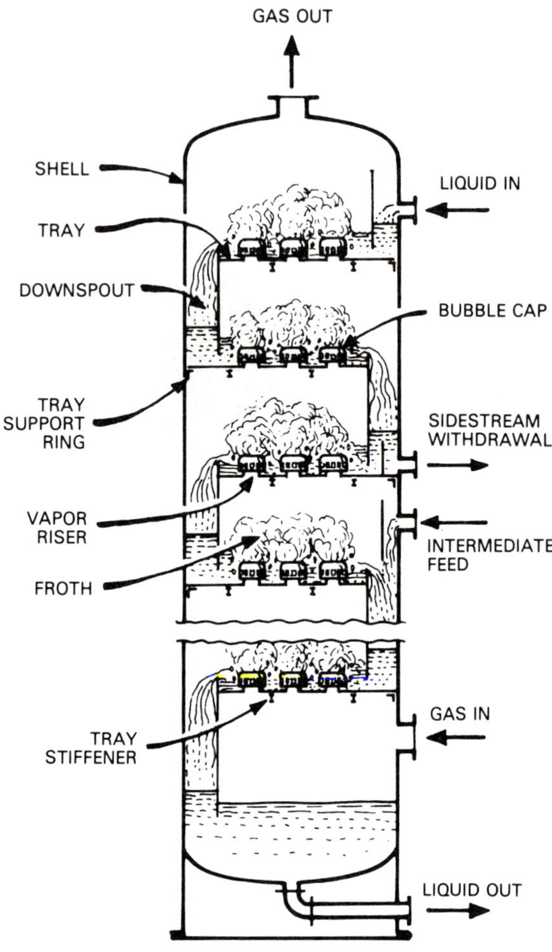

Figure B.3. Bubble-cap tray tower

Absorber and stripper units

Absorbers and strippers are built much alike. The object of the stripper is to remove something from the liquid stream with gas; in the absorber, liquids are removed from the gas. These may be packed-type (fig. B.2) or tray-type (fig. B.3) towers. The tray type may have bubble-cap trays, float-valve trays, or sieve trays, depending upon the process under consideration.

Fractionators and stabilizers (fig. B.4) are very similar to absorbers and strippers except that the feed point is either near the top or at

Figure B.4. Fractionator and stabilizer towers

midpoint, and a liquid or a liquid-gas feed is made to the tower. Overhead condensers and reboilers are used to control the top and bottom operating temperatures of fractionators.

Heat exchangers in gas processing plants are usually shell-and-tube water-cooled units or air-cooled exchangers. Air-cooled exchangers are being used more and more today because of lower cost, elimination of water supply and treatment problems, and ease of maintenance. Shell-and-tube exchangers require more periodic cleaning to eliminate fouling. In some cases, chemical cleaning is required and even periodic bundle replacement. Air-cooled exchangers are usually provided with temperature-controlled louvers so that the outlet process temperature can be carefully controlled. Recent use of recycle air ducting on air coolers has provided increased flexibility in adapting air-cooled exchangers to processing plants for northern climates. Additional temperature control can be obtained with air coolers by using variable-speed fan motors, adjustable-pitch fans, and variable-speed fan drives. Heat in the process is supplied through the use of indirect heaters, hot-oil heaters, salt-bath heaters, and direct-fired heaters.

Glossary

A

abandonment pressure *n:* the average reservoir pressure at which an amount of gas insufficient to permit continued economic operation of a producing gas well is expelled.

absolute open flow *n:* the maximum flow rate that a well could theoretically deliver with zero pressure at the sand face.

absolute pressure *n:* total pressure measured from an absolute vacuum. It equals the sum of the gauge pressure and the atmospheric pressure corresponding to the barometer (expressed in pounds per square inch).

absolute temperature scale *n:* a scale of temperature measurement in which zero degrees is absolute zero. On the Rankine absolute temperature scale, in which degrees correspond to degrees Fahrenheit, water freezes at 492 degrees and boils at 672 degrees. On the kelvin absolute temperature scale, in which degrees correspond to degrees Celsius, water freezes at 273 degrees and boils at 373 degrees. See *absolute zero.*

absolute zero *n:* a hypothetical temperature at which there is a total absence of heat. Since heat is a result of energy caused by molecular motion, there is no motion of molecules with respect to each other at absolute zero.

absorber *n:* 1. a vertical, cylindrical vessel that recovers heavier hydrocarbons from a mixture of predominantly lighter hydrocarbons. Also called absorption tower. 2. a vessel in which gas is dehydrated by being bubbled through glycol.

absorption *n:* the process of sucking up; taking in and making part of an existing whole. Compare *adsorption.*

acid gas *n:* a gas that forms an acid when mixed with water. In petroleum production and processing, the most common acid gases are hydrogen sulfide and carbon dioxide. They both cause corrosion, and hydrogen sulfide is very poisonous.

adiabatic compression *n:* the compression of air or gas that exists when no heat is transferred between the air or gas and surrounding bodies (such as the cylinders and pistons in a compressor). It is characterized by an increase in temperature during compression and a decrease in temperature during expansion.

adsorption *n:* the adhesion of a thin film of gas or liquid to the surface of a solid. Liquid hydrocarbons are recovered from natural gas by passing the gas through activated charcoal, silica gel, or other solids, which extract the heavier hydrocarbons. Steam treatment of the solid removes the adsorbed hydrocarbons, which are then collected and recondensed. The adsorption process is also used to remove water vapor from air or natural gas. Compare *absorption.*

alkanolamine *n:* a chemical family of specific organic compounds, including monoethanolamine (MEA), diethanolamine (DEA), and triethanolamine (TEA). These chemicals, and proprietary mixtures containing them and other amines, are used extensively for the removal of hydrogen sulfide and carbon dioxide from other gases and are particularly adapted for obtaining the low acid gas residuals that are usually specified by pipelines.

alkanolamine process *n:* a continuous-operation liquid process for acid gas removal from natural gas, using chemical absorption with subsequent heat addition to strip the acid gas components from the absorbent solution.

Amagat-Leduc rule *n:* the rule stating that the volume occupied by a mixture of gases equals the sum of the volumes each gas would occupy at the pressure and temperature of the mixture.

annular space *n:* 1. the space surrounding a cylindrical object within a cylinder. 2. the space around a pipe in a wellbore, the outer wall of which may be the wall of either the borehole or the casing; sometimes termed the annulus.

annulus *n:* also called annular space. See *annular space.*

anticline *n:* an arched, inverted-trough configuration of folded rock layers.

API gravity *n:* the measure of the density or gravity of liquid petroleum products in the United States, derived from specific gravity in accordance with the following equation:

$$\text{API gravity} = \frac{141.5}{\text{specific gravity}} - 131.5.$$

API gravity is expressed in degrees, 10°API being equivalent to 1.0, the specific gravity of water.

aromatic hydrocarbons *n pl:* hydrocarbons derived from or containing a benzene ring. Many have an odor. Single-ring aromatic hydrocarbons are the benzene series (benzene, ethylbenzenes, and toluene). Aromatic hydrocarbons also include naphthalene and anthracene.

associated gas *n:* natural gas that overlies and contacts crude oil in a reservoir. Where reservoir conditions are such that the production of associated gas does not substantially affect the recovery of crude oil in the reservoir, such gas may be reclassified as nonassociated gas by a regulatory agency. Also called associated free gas.

Avogadro's law *n:* the law that states that under the same conditions of pressure and temperature, equal volumes of all gases contain equal numbers of molecules. Also called Avogadro's hypothesis.

B

babbitt *n:* metal alloy, either tin-based or lead-based, primarily used in friction bearings.

back-pressure *n:* 1. the pressure maintained on equipment or systems through which a fluid flows. 2. in reference to engines, a term used to describe the resistance to the flow of exhaust gas through the exhaust pipe.

base pressure *n:* the pressure to which gas volumes are calculated, regardless of the pressure at which they are measured. The standard base pressure for gas volume calculations in the United States varies from state to state. For example, a standard cubic foot of gas in Texas is not the same as a standard cubic foot of gas in Louisiana.

bhp *abbr:* brake horsepower.

BHP *abbr:* bottomhole pressure.

block valve *n:* a valve that completely shuts off flow of a fluid. It is usually either completely open or completely closed.

blowout *n:* an uncontrolled flow of gas, oil, or other well fluids into the atmosphere. A blowout, or gusher, can occur when formation pressure exceeds the pressure applied to it by the column of drilling fluid. A kick warns of the possibility of a blowout.

bottomhole pressure *n:* the pressure at the bottom of a borehole. It is caused by the hydrostatic pressure of the wellbore fluid and, sometimes, any back-pressure held at the surface, as when the well is shut in with blowout preventers. When mud is being circulated, bottomhole pressure is the hydrostatic pressure plus the remaining circulating pressure required to move the mud up the annulus.

Bourdon tube *n:* a flattened metal tube bent in a curve, which tends to staighten when pressure is applied internally. By the movements of an indicator over a circular scale, a Bourdon tube indicates the pressure applied.

brake horsepower *n:* the power produced by an engine as it is measured by the force applied to a friction brake or by an absorption dynamometer applied to the shaft or the flywheel.

British thermal unit *n:* a measure of heat energy equivalent to the amount of heat needed to raise 1 pound of water 1 degree Fahrenheit.

Btu *abbr:* British thermal unit.

bubble point *n:* 1. the temperature and pressure at which part of a liquid begins to convert to gas. For example, if a certain volume of liquid is held at constant pressure, but its temperature is increased, a point is reached when bubbles of gas begin to form in the liquid. That is the bubble point. Similarly, if a certain volume of liquid is held at a constant temperature but the pressure is reduced, the point at which gas begins to form is the bubble point. Compare *dew point.* 2. the temperature and pressure at which gas, held in solution in crude oil, breaks out of solution as free gas.

buildup test *n:* a test in which a well is shut in for a prescribed period of time and a bottomhole pressure bomb run in the well to record the pressure. From this data and from knowledge of pressures in a nearby well, the effective drainage radius or the presence of permeability barriers or other production deterrents surrounding the wellbore can be estimated.

C

calorimeter *n:* an apparatus used to determine the heating value of a combustible material.

casinghead gas *n:* gas produced with oil.

Celsius scale *n:* the metric scale of temperature measurement used universally by scientists. On this scale, 0 degrees represents the freezing point of water and 100 degrees its boiling point at a barometric pressure of 760 mm. Degrees Celsius are converted to degrees Fahrenheit by using the following equation:

$$°F = 9/5\ (°C) + 32.$$

The Celsius scale was formerly called the centrigrade scale; now, however, the term *Celsius* is preferred in the International System of Units (SI).

centrifugal compressor *n:* a compressor in which the flow of gas to be compressed is moved away from the center rapidly, usually by a series of blades, or turbines. It is a continuous-flow compressor with a low pressure ratio, often used to transmit gas through a pipeline. Gas passing through the compressor contacts a rotating impeller, from which it is discharged into a diffuser, where its velocity is slowed and its kinetic energy changed to static pressure. Centrifugal compressors are nonpositive-displacement machines, often arranged in series on a line to achieve multistage compression.

charcoal test *n:* a test standardized by the American Gas Association and the Gas Processors Association for determining the natural gasoline content of a given natural gas. The gasoline is adsorbed from the gas on activated charcoal and then recovered by distillation. The test is described in Testing Code 101-43, a joint AGA and GPA publication.

choke *n:* a device with an orifice installed in a line to restrict the flow of fluids. Surface chokes are part of the Christmas tree on a well and contain a choke nipple, or bean, with a small-diameter bore that serves to restrict the flow. Chokes are also used to control the rate of flow of the drilling mud out of the hole when the well is closed in with the blowout preventer and a kick is being circulated out of the hole.

choke line *n:* a pipe attached to the blowout preventer stack, out of which kick fluids and mud can be pumped to the choke manifold when a blowout preventer is closed in on a kick.

Christmas tree *n:* the control valves, pressure gauges, and chokes assembled at the top of a well to control the flow of oil and gas after the well has been drilled and completed.

compressibility factor *n:* a factor, usually expressed as *Z,* which gives the ratio of the actual volume of gas at a given temperature and pressure to the volume of gas when calculated by the ideal gas law without any consideration of the compressibility factor. The compressibility factor can be inserted into the ideal gas law for the purpose of taking into account the deviation of true gases from ideal gas behavior.

compression ratio *n:* in a compressor, the ratio of the absolute discharge pressure to the absolute suction pressure. This must not be confused with the compression ratio used in power cylinders of internal-combustion engines, which is the displacement plus the clearance divided by the clearance.

compression test *n:* a type of testing used extensively on casinghead gas that has a $C_4{}^+$ content of above 0.5–1.0 gallons per thousand cubic feet to determine the liquid hydrocarbon content of the gas.

compressor *n:* a device that raises the pressure of a compressible fluid such as air or gas. Compressors create a pressure differential to move or compress a vapor or a gas, consuming power in the process. They may be positive-displacement compressors or nonpositive-displacement compressors.

compressor clearance *n:* the ratio of the volume remaining in a compressor cylinder at the end of a compression stroke to the volume displaced by one stroke of the piston. The ratio is usually expressed in percent.

condensate *n:* a light hydrocarbon liquid obtained by condensation of hydrocarbon vapors. It consists of varying proportions of butane, propane, pentane, and heavier fractions, with little or no methane or ethane.

condensate liquids *n pl:* hydrocarbons that are gaseous in the reservoir but will separate out in liquid form at the pressures and temperatures at which separators normally operate; sometimes called distillate.

condensate ratio *n:* the ratio of the volume of liquid produced to the volume of residue gas produced and usually expressed in barrels per million cubic feet.

condensation *n:* the process by which vapors are converted into liquids, chiefly accomplished by cooling the vapors. Condensation is often the cause of water appearing in fuels.

connate water *n:* water retained in the pore spaces, or interstices, of a formation from the time the formation was created. Compare *interstitial water.*

contactor *n:* a vessel or piece of equipment in which two or more substances are brought together.

controller *n:* an electric device used for governing the power that goes to an apparatus to which it is connected.

convergence pressure *n:* the pressure at a given temperature for a hydrocarbon system of fixed composition at which the vapor-liquid equilibria values of the various components in the system become or tend to become unity. The convergence pressure is used to adjust vapor-liquid equilibria values to the particular system under consideration.

corrosion coupon *n:* a metal strip inserted into a system to monitor corrosion rate and to indicate corrosion-inhibitor effectiveness.

corrosion inhibitor *n:* a chemical substance that minimizes or prevents corrosion from occurring in metal equipment.

cracking *n:* the process of breaking down large chemical compounds into smaller compounds under the influence of heat or catalysts. In petroleum refining, the two major types of cracking are *thermal cracking* and *catalytic cracking.*

crankshaft *n:* a rotating shaft to which connecting rods are attached. It changes up and down (reciprocating) motion into circular (rotary) motion.

critical pressure *n:* the pressure needed to condense a vapor at its critical temperature.

critical temperature *n:* the highest temperature at which a substance can be separated into two fluid phases—liquid and vapor. Above the critical temperature, a gas cannot be liquefied by pressure alone.

cyclic compound *n:* a compound that contains a ring of atoms.

cylinder head *n:* the device used to seal the top of a cylinder. In modern drilling rig engines, it also houses the valves and has exhaust passages. In four-cycle operation, the cylinder head also has intake passages.

D

Dalton's law *n:* the law that the pressure of a mixture of gases is equal to the sum of the partial pressures of the gases of which it is composed. Also called law of partial pressures.

dehydration *n:* the removal of water or water vapor from gas or oil. See *dehydrator.*

dehydrator *n:* equipment or apparatus for effecting dehydration.

deliverability test *n:* a type of test for either an oilwell or a gas well to determine the actual flow rate.

density *n:* the mass or weight of a substance per unit volume. For instance, the density of a drilling mud may be 10 pounds per gallon (ppg), 74.8 pounds per cubic foot (lb/ft^3), or 1 198.2 kilograms per cubic metre (kg/m^3). Specific gravity, relative density, and API gravity are other units of density.

depletion drive *n:* also called gas drive. See *gas drive.*

desiccant *n:* a substance able to remove water from another substance with which it is in contact. It may be liquid (as triethylene glycol) or solid (as silica gel).

dew point *n:* the temperature and pressure at which a liquid begins to condense out of a gas. For example, if a constant pressure is held on a certain volume of gas but the temperature is reduced, a point is reached at which droplets of liquid condense out of the gas. That point is the dew point of the gas at that pressure. Similarly, if a constant temperature is maintained on a volume of gas but the pressure is increased, the point at which liquid begins to condense out is the dew point at that temperature. Compare *bubble point.*

differential *n:* the difference in quantity or degree between two measurements or units. For example, the pressure differential across a choke is the variation between the pressure on one side and that on the other.

displacement meter *n:* a meter in which a piston is actuated by the pressure of a measured volume of liquid, and the volume swept by the piston is equal to the volume of the liquid recorded.

dolomitization *n:* the shrinking of the solid volume of rock as limestone turns to dolomite; the conversion of limestone to dolomite rock by replacement of a portion of the calcium carbonate with magnesium carbonate.

dome *n:* a geologic structure resembling an inverted bowl; a short anticline that dips or plunges on all sides.

drawdown *n:* 1. the difference between static and flowing bottomhole pressures. 2. the distance between the static level and the pumping level of the fluid in the annulus of a pumping well.

dry gas *n:* 1. gas whose water content has been reduced by a dehydration process. 2. gas containing few or no hydrocarbons commercially recoverable as liquid product. Also called lean gas.

E

effective porosity *n:* the percentage of the bulk volume of a rock sample that is composed of interconnected pore spaces that allow the passage of fluids through the sample. See *porosity.*

emulsion *n:* a mixture in which one liquid, termed the dispersed phase, is uniformly distributed (usually as minute globules) in another liquid, called the continuous phase or dispersion medium. In an oil-water emulsion, the oil is the dispersed phase and the water the dispersion medium; in a water-oil emulsion, the reverse holds. A typical product of oilwells, water-oil emulsion is also used as a drilling fluid.

energy *n:* the capability of a body for doing work. Potential energy is this capability due to the position or state of the body. Kinetic energy is the capability due to the motion of the body.

equilibrium constant *n:* See *vapor-liquid equilibrium ratio.*

F

Fahrenheit scale *n:* a temperature scale devised by Gabriel Fahrenheit, in which 32 degrees represents the freezing point and 212 degrees the boiling point of water at standard sea-level pressure. Fahrenheit degrees may be converted to Celsius degrees by using the following formula:

$$°C = 5/9\ (°F - 32)$$

fault *n:* a break in the earth's crust along which rocks on one side have been displaced (upward, downward, or laterally) relative to those on the other side.

field processing *n:* the processing of oil and gas in the field before delivery to a major refinery or gas plant, including separation of oil from gas, separation of water from oil and from gas, and removal of liquid hydrocarbons.

flash tank *n:* a vessel used for separating the liquid phase from the gaseous phase formed from a rise in temperature and/or a reduction of pressure on the flowing stream.

force *n:* that which causes, changes, or stops the motion of a body.

formation fracturing *n:* a method of stimulating production by opening new flow channels in the rock surrounding a production well. In hydraulic fracturing, a fluid such as water, oil, alcohol, or dilute hydrochloric acid is pumped downward at high pressure through tubing or drill pipe and forced into the perforations in the casing. The fluid enters the formation and parts or fractures it. Sand grains, aluminum pellets, glass beads, or similar materials are carried in suspension by the fluid into the fractures. These are called propping agents or proppants. When the pressure is released at the surface, the fractures partially close on the proppants, leaving channels for oil to flow through them to the well. In explosive fracturing, explosives are used to fracture the formation. At the instant of detonation, the explosion also furnishes a source of high-pressure gas to force fluid into the formation. The rubble resulting from the explosion prevents fracture healing, making the use of proppants unnecessary. Formation fracturing is often called a frac job.

four-stroke/cycle engine *n:* an engine in which the piston moves from top dead center to bottom dead center two times to complete a cycle of events. The crankshaft must make two complete revolutions, or 720 degrees.

fractional analysis *n:* a test for the composition of gas or two-phase gas-condensate streams. The analysis generally shows not only the composition in percent of each hydrocarbon present through hexanes or heptanes but also the gallons per thousand cubic feet of liquids by component and the heating value of the gas.

fractionate *v:* to separate single fractions from a mixture of hydrocarbon fluids, usually by distillation.

fractionating column *n:* the vessel or tower in a gas plant in which fractionation occurs. See *fractionate.*

fracturing *n:* shortened form of formation fracturing. See *formation fracturing.*

G

gas-cap drive *n:* drive energy supplied naturally (as a reservoir is produced) by the expansion of gas in a cap overlying the oil in the reservoir. See *reservoir drive mechanism.*

gas drive *n:* the use of the energy that arises from the expansion of compressed gas in a reservoir to move crude oil to a wellbore. See *reservoir drive mechanism.*

gasoline *n:* a volatile, flammable liquid hydrocarbon refined from crude oils and used universally as a fuel for internal-combustion, spark-ignition engines.

gasoline plant *n:* also called a natural gas processing plant, a term that is preferred because it distinguishes the plant from a unit that makes gasoline within an oil refinery. See *natural gas processing plant.*

gas processing plant *n:* also called natural gas processing plant. See *natural gas processing plant.*

gas turbine *n:* an engine in which gas, under pressure or formed by combustion, is directed against a series of turbine blades. The energy in the expanding gas is converted into rotary motion.

gate valve *n:* a type of valve. See *valve.*

gathering system *n:* the pipelines and other equipment needed to transport oil, gas, or both from wells to a central point—the gathering station—where there is the accessory equipment required to deliver a clean and salable product to the market or to another pipeline. An oil gathering system includes oil and gas separators, emulsion treaters, gathering tanks, and similar equipment. A gas gathering system includes regulators, compressors, dehydrators, and associated equipment.

gauge pressure *n:* the amount of pressure exerted on the interior walls of a vessel by the fluid contained in it (as indicated by a pressure gauge); it is expressed in psig (pounds per square inch gauge) or in kilopascals. Gauge pressure plus atmospheric pressure equals absolute pressure. See *absolute pressure.*

glycol *n:* a group of compounds used to dehydrate gaseous or liquid hydrocarbons or to inhibit the formation of hydrates. Commonly used glycols are ethylene glycol, diethylene glycol, and triethylene glycol.

glycol/amine process *n:* a process that uses a solution comprised of 10 to 30 weight percent monoethanolamine, 45 to 85 percent glycol, and 5 to 25 percent water for the simultaneous removal of water vapor, H_2S, and CO_2 from gas streams.

H

heater *n:* a container or vessel enclosing an arrangement of tubes and a firebox in which an emulsion is heated before further treating.

heat of vaporization *n:* the quantity of energy required to evaporate 1 mole of a liquid at constant pressure and temperature.

hydrate *n:* a hydrocarbon and water compound that is formed under reduced temperature and pressure in gathering, compression, and transmission facilities for gas. Hydrates often accumulate in troublesome amounts and impede fluid flow. They resemble snow or ice. *v:* to enlarge by taking water on or in.

hydrocarbon pore volume *n:* the volume of the pore space in a reservoir that is occupied by oil, natural gas, or other hydrocarbons (including nonhydrocarbon impurities). It may be expressed in acre-feet, barrels, or cubic feet.

hydrocarbons *n pl:* organic compounds of hydrogen and carbon, whose densities, boiling points, and freezing points increase as their molecular weights increase. Although composed of only two elements, hydrocarbons exist in a variety of compounds because of the strong affinity of the carbon atom for other atoms and for itself. The smallest molecules of hydrocarbons are gaseous; the largest are solids. Petroleum is a mixture of many different hydrocarbons.

hydrometer *n:* an instrument with a graduated stem, used to determine the gravity of liquids. The liquid to be measured is placed in a cylinder, and the hydrometer dropped into it. It floats at a certain level in the liquid (high if the liquid is light, low if it is heavy), and the stem markings indicate the gravity of the liquid.

I

ideal gas *n:* 1. a gas whose molecules are infinitely small and exert no force on each other. 2. a gas that obeys Boyle's law and Joule's law. Also called a perfect gas.

ideal gas law *n:* the equation of state of an ideal gas, showing a close approximation to real gases at sufficiently high temperature and low pressures; that is—

$$PV = RT$$

where

P = pressure
V = volume per mole of gas
T = temperature
R = gas constant.

impeller *n:* a set of mounted blades used to impart motion to a fluid (e.g., the rotor of a centrifugal pump).

indirect heater *n:* apparatus or equipment in which heat from a primary source, usually the combustion of fuel, is transferred to a fluid or solid which acts as the heating medium.

international system of units *n:* a system of units of measurement based on the metric system, adopted and described by the Eleventh General Conference on Weights and Measures. It provides an international standard of measurement to be followed when certain customary units, both U.S. and metric, are eventually phased out of international trade operations. The symbol *SI* (Le Système International d'Unités) designates the system, which involves seven base units: (1) metre for length, (2) kilogram for mass, (3) second for time, (4) kelvin for temperature, (5) ampere for electric current, (6) candela for luminous intensity, and (7) mole for amount of substance. From these units, others are derived without introducing numerical factors.

interstitial water *n:* water contained in the interstices, or pores, of reservoir rock. In reservoir engineering, it is synonymous with connate water. Compare *connate water.*

iron sponge process *n:* a method for removing small concentrations of hydrogen sulfide from natural gas by passing the gas over a bed of wood shavings which have been impregnated with a form of iron oxide. The impregnated wood shavings are called iron sponge. The hydrogen sulfide reacts with the iron oxide, forming iron sulfide and water.

isochronal test *n:* a short-time back-pressure test for low-permeability reservoirs that otherwise require excessively long times for pressure stabilization when wells are shut in.

isothermal compression *n:* the compression of air or gas that exists when the interchange of heat between the air or gas and surrounding bodies (i.e., cylinders or pistons) takes place at a rate exactly sufficient to maintain the air or gas at constant temperature as the pressure increases.

K

kelvin temperature scale *n:* a temperature scale with the degree interval of the Celsius scale and the zero point at absolute zero. On the kelvin scale, water freezes at 273 degrees and boils at 373 degrees. See *absolute temperature scale.*

kill *v:* 1. in drilling, to prevent a threatened blowout by taking suitable preventive measures (e.g., to shut in the well with the blowout preventers, circulate the kick out, and increase the weight of the drilling mud). 2. in production, to stop a well from producing oil and gas so that reconditioning of the well can proceed. Production is stopped by circulating a kill fluid into the hole.

kill line *n:* a pipe attached to the blowout preventer stack, into which mud or cement can be pumped to overcome the pressure of a kick; sometimes used when normal kill procedures (circulating kill fluids down the drill stem) are not sufficient.

kill string *n:* small-diameter tubing that is used inside production tubing for continuous injection of specialized fluids such as corrosion inhibitors or kill fluids. Sometimes used to refer to the drill string through which kill fluids are circulated in drilling.

kinetic energy *n:* energy possessed by a body because of its motion.

L

liquefied petroleum gas *n:* a mixture of heavier, gaseous, paraffinic hydrocarbons, principally butane and propane. These gases, easily liquefied at moderate pressure, may be transported as liquids but converted to gases on release of the pressure. Thus, liquefied petroleum gas is a portable source of thermal energy that finds wide application in areas where it is impractical to distribute natural gas. It is also used as a fuel for internal-combustion engines and has many industrial and domestic uses. Principal sources are natural and refinery gas, from which the liquefied petroleum gases are separated by fractionation.

M

manometer *n:* a U-shaped piece of glass tubing containing a liquid (usually water or mercury) that is used to measure the pressure of gases or liquids. When pressure is applied, the liquid level in one arm rises while the level in the other drops. A set of calibrated markings beside one of the arms permits a pressure reading to be taken, usually in inches or millimetres.

mass-transfer zone *n:* the depth of solid desiccant bed, in a solid desiccant dehydration system, from saturation to initial adsorption.

matrix *n:* in rock, the fine-grained material between larger grains, in which the larger grains are embedded. A rock matrix may be composed of fine sediments, crystals, clay, or other substances.

meter chart *n:* a circular chart of special paper that shows the range of differential pressure and static pressure and is marked by the recording pens of a flow meter.

meter tube *n:* an important part of the primary element of an orifice meter installation that must create a known flow pattern for the fluid as it reaches the plate. It is the straight upstream pipe of the same size between the orifice and the nearest pipe fitting and the similar downstream pipe between the orifice and the nearest pipe fitting.

methanol (methyl alcohol) *n:* the lightest alcohol, having the chemical formula CH_3OH. Also called wood alcohol.

Mcf *abbr:* 1,000 cubic feet of gas, commonly used to express the volume of gas produced, transmitted, or consumed in a given period.

Mcf/d *abbr:* 1,000 ft^3 of gas per day.

mercaptan *n:* a compound chemically similar to alcohol, with sulfur replacing oxygen in the chemical structure. Many mercaptans have an offensive odor and are used as odorants in natural gas.

micron *n:* one-millionth of a metre; a metric unit of measure of length equal to 0.001 mm.

MMcf *abbr:* million cubic feet; a common unit of measurement for large quantities of gas.

MMscf *abbr:* million standard cubic feet. The standard referred to is usually 60°F and 1 atmosphere (14.7 psi) of pressure but varies from state to state.

MMscf/d *abbr:* million standard cubic feet per day.

mol *sym:* mole.

mole *n:* the fundamental unit of mass of a substance. Its symbol is mol. A mole of any substance is the number of grams or pounds indicated by its molecular weight. For example, water, H_2O, has a molecular weight of approximately 18. Therefore, a gram-mole of water is 18 grams of water; a pound-mole of water is 18 pounds of water.

molecular sieves *n pl:* synthetic zeolites packaged in bead or pellet form for (1) use in recovering contaminants or impurities from liquid and vapor product streams by selective adsorption and (2) use as a catalyst.

N

natural gas *n:* a highly compressible, highly expansible mixture of hydrocarbons having a low specific gravity and occurring naturally in gaseous form. Besides hydrocarbon gases, natural gas may contain appreciable quantities of nitrogen, helium, carbon dioxide, hydrogen sulfide, and water vapor. Although gaseous at normal temperatures and pressures, the gases comprising the mixture that is natural gas are variable in form and may be found either as gases or as liquids under suitable conditions of temperature and pressure.

natural gas liquids *n:* those hydrocarbons liquefied at the surface in field facilities or in gas processing plants. Natural gas liquids include propane, butane, and natural gasoline.

natural gas plant *n:* also called natural gas processing plant. See *natural gas processing plant.*

natural gas processing plant *n:* an installation in which natural gas is processed for recovery of natural gas liquids, the heavier hydrocarbon components of natural gas, including liquefied petroleum gases (such as butane and propane), and natural gasoline. The modern and preferred term for gas processing plant, natural gas plant, gasoline plant, and others.

nonassociated gas *n:* gas in a reservoir that contains no oil.

O

oil and gas separator *n:* an item of production equipment used to separate liquid components of the well stream from gaseous elements. Separators are either vertical or horizontal and either cylindrical or spherical in shape. Separation is accomplished principally by gravity, the heavier liquids falling to the bottom and the gas rising to the top. A float valve or other liquid-level control regulates the level of oil in the bottom of the separator.

open flow test *n:* a test made to determine the volume of gas that will flow from a well during a given time span when all surface control valves are wide open.

orifice *n:* an opening of a measured diameter, used for measuring the flow of fluid through a pipe or delivering a given amount of fluid through a fuel nozzle. In measuring the flow of fluid through a pipe, the orifice must be of smaller diameter than the pipe diameter; it is placed in an orifice plate held by an orifice fitting.

orifice meter *n:* an instrument used to measure the flow of fluid through a pipe. The orifice meter is an inferential device, which measures and records the pressure differential created by the passage of a fluid through an orifice of critical diameter placed in the line. The rate of flow is calculated from the differential pressure and the static, or line, pressure and other factors such as the temperature and density of the fluid, the size of the pipe, and the size of the orifice.

orifice plate *n:* a sheet of metal, usually circular, in which a hole of specific size is made for use in an orifice fitting.

P

packer *n:* a piece of downhole equipment, consisting of a sealing device, a holding or setting device, and an inside passage for fluids, used to block the flow of fluids through the annular space between the tubing and the wall of the wellbore by sealing off the space between them. It is usually made up in the tubing string some distance above the producing zone. A sealing element expands to prevent fluid flow except through the inside bore of the packer and into the tubing. Packers are classified according to configuration, use, and method of setting and whether or not they are retrievable (that is, whether they can be removed when necessary, or whether they must be milled or drilled out and thus destroyed).

paraffin *n:* a hydrocarbon having the formula C_nH_{2n+2} (e.g., methane, CH_4; ethane, C_2H_6). Heavier paraffin hydrocarbons (i.e., $C_{18}H_{38}$ and heavier) form a waxlike substance that is called paraffin. These heavier paraffins often accumulate on the walls of tubing and other production equipment, restricting or stopping the flow of desirable lighter paraffins.

partial pressure *n:* the pressure exerted by one specific component of a gaseous mixture.

permeability *n:* 1. a measure of the ease with which a fluid flows through the connecting pore spaces of rock or cement. The unit of measurement is the millidarcy. 2. fluid conductivity of a porous medium. 3. ability of a fluid to flow within the interconnected pore network of a porous medium.

pH *abbr:* an indicator of the acidity or alkalinity of a substance or solution, represented on a scale of 0–14; 0–7 being acidic, 7–14 basic. These values are based on hydrogen ion content and activity.

piston displacement *n:* the actual volume displaced by a piston in a cylinder as it moves from the beginning of the compression stroke to the end of it.

pneumatic control *n:* a control valve that is actuated by air. Several pneumatic controls are used on drilling rigs to actuate rig components (clutches, hoists, engines, pumps, etc.)

pore *n:* an opening or space within a rock or mass of rocks, usually small and often filled with some fluid (water, oil, gas, or all three).

pore volume *n:* the total volume of pore space in a reservoir formation.

porosity *n:* 1. the condition of being porous (such as a rock formation). 2. the ratio of the volume of empty space to the volume of solid rock in a formation, indicating how much fluid a rock can hold.

power *n:* the rate of doing work, or the amount of work done in a specific unit of time. In mechanics, it is calculated in foot-pounds per minute. One horsepower equals 33,000 foot-pounds per minute.

pressure base *n:* See *base pressure.*

pressure depletion *n:* the method of production of a gas reservoir that is not associated with a water drive. It is the process in which gas is removed and reservoir pressure declines until all the recoverable gas has been expelled.

pressure gradient *n:* a scale of pressure differences in which there is a uniform variation of pressure from point to point. For example, the pressure gradient of a column of water is about 0.433 psi/ft (9.794 kPa/m) of vertical elevation. The normal pressure gradient in a formation is equivalent to the pressure exerted at any given depth by a column of 10 percent salt water extending from that depth to the surface (0.465 psi/ft or 10.518 kPa/m).

prime mover *n:* an internal-combustion engine or a turbine that is the source of power for driving a machine or machines.

pseudocritical properties *n pl:* empirical values for the critical properties (such as temperature, pressure, and volume) of a chemical system made up of multiple components.

psia *abbr:* pounds per square inch absolute. Psia is equal to the gauge pressure plus the pressure of the atmosphere at that point.

psig *abbr:* pounds per square inch gauge.

pyrometer *n:* an instrument for measuring temperatures, especially those above the range of mercury thermometers.

R

Rankine temperature scale *n:* a temperature scale with the degree interval of the Fahrenheit scale and the zero point at absolute zero. On the Rankine scale, water freezes at 491.60° and boils at 671.69°. See *absolute temperature scale.*

reciprocating compressor *n:* a type of compressor that has pistons moving back and forth in cylinders, suction valves, and discharge valves; a positive-displacement compressor. Reciprocating compressors are used extensively in the transmission of natural gas through pipelines.

reflux *n:* in the distillation process, that part of the condensed overhead stream that is returned to the fractionating column as a source of cooling.

regeneration gas *n:* wet gas that has been heated in a regeneration gas heater to temperatures of 400° to 460°F in a solid desiccant dehydration system. The gas is passed through a saturated adsorber tower to dry the tower and remove the previously adsorbed water.

regulator *n:* a device that reduces the pressure or volume of a fluid flowing in a line and maintains the pressure or volume at a specified level.

reserves *n pl:* the unproduced but recoverable oil or gas in a formation that has been proved by production.

reservoir *n:* a subsurface, porous, permeable rock body in which oil and/or gas has accumulated. Most reservoir rocks are limestones, dolomites, sandstones, or a combination of these. The three basic types of hydrocarbon reservoirs are oil, gas, and condensate. An oil reservoir generally contains three fluids—gas, oil, and water—with oil the dominant product. In the typical oil reservoir, these fluids become vertically segregated because of their different densities. Gas, the lightest, occupies the upper part of the reservoir rocks; water, the lower part; and oil, the intermediate section. In addition to its occurrence as a cap or in solution, gas may accumulate independently of the oil; if so, the reservoir is called a gas reservoir. Associated with the gas, in most instances, are salt water and some oil. In a condensate reservoir, the hydrocarbons may exist as a gas, but, when brought to the surface, some of the heavier ones condense to a liquid.

reservoir drive mechanism *n:* the process in which reservoir fluids are caused to flow out of the reservoir rock and into a wellbore by natural energy. Gas drives depend on the fact that, as the reservoir is produced, pressure is reduced, allowing the gas to expand and provide the principal driving energy. Water drive reservoirs depend on water and rock expansion to force the hydrocarbons out of the reservoir and into the wellbore.

reservoir pressure *n:* the average pressure within the reservoir at any given time. Determination of this value is best made by bottomhole pressure measurements with adequate shut-in time. If a shut-in period long enough for the reservoir pressure to stabilize is impractical, then various techniques of analysis by pressure buildup or drawdown tests are available to determine static reservoir pressure.

reservoir temperature *n:* the average temperature within the reservoir, measured during logging, drill stem testing, or bottomhole pressure testing using a bottomhole temperature recorder.

residual gas saturation *n:* the portion of the hydrocarbons that cannot be removed by ordinary producing mechanisms when a porous reservoir has been saturated with hydrocarbons. This value is usually a specific percentage of the pore volume. In the case of gas, the volume, measured at standard conditions, that is retained in a reservoir as residual gas saturation is an inverse function of the pressure, due to the effect of the gas laws.

retrograde reservoir *n:* a reservoir in which the pressure is high and the hydrocarbon content is completely in a gaseous or supercritical phase at initial conditions, and as the pressure drops because of production, the heavier hydrocarbon components condense, forming liquids within the reservoir. Such action is the retrograde phenomenon. If the reservoir pressure is completely depleted, only a small portion of these liquids will revaporize and be recovered.

rotor *n:* a device with vanelike blades attached to a shaft; the device turns or rotates when the vanes are struck by a fluid directed there by a stator.

S

scrubber *n:* 1. a vessel through which fluids are passed to remove dirt, other foreign matter, or an undesired component of the fluid. 2. a vessel with or without internals used to separate entrained liquids or solids from gas. It may be used to protect downstream rotating equipment or to recover valuable liquids from a gas or vapor originating upstream.

separator *n:* a cylindrical or spherical vessel used to isolate the components in mixed streams of fluids. See *oil and gas separator.*

shut-in *adj:* shut off to prevent flow. Said of a well, plant, pump, and so forth, when valves are closed at both inlet and outlet.

SI *abbr:* 1. shut in; used in drilling reports. 2. Système International. See *international system of units.*

single-phase flow *n:* the flow of only one material—a gas or a liquid—in a pipeline.

slugging *n:* the accumulation of a liquid (water, oil, or condensate) in a gas line.

solution gas *n:* lighter hydrocarbons that exist as a liquid under reservoir conditions but that effervesce as gas when pressure is released during production.

solution-gas drive *n:* a source of natural reservoir energy, in which the solution gas coming out of the oil expands to force the oil into the wellbore.

sorber *n:* a vessel for absorption, adsorption, or desorption.

sour *adj:* containing or caused by hydrogen sulfide or another acid gas (e.g., sour crude, sour gas, sour corrosion).

specific gravity *n:* the ratio of the weight of a given volume of a substance at a given temperature to the weight of an equal volume of a standard substance at the same temperature. For example, if 1 cubic inch of water at 39°F weighs 1 unit and 1 cubic inch of another solid or liquid at 39° weighs 0.95 unit, then the specific gravity of the substance is 0.95. In determining the specific gravity of gases, the comparison is made with the standard of air or hydrogen.

specific heat *n:* the amount of heat required to cause a unit increase in temperature in a unit mass of a substance, expressed as numerically equal to the number of calories needed to raise the temperature of 1 g of a substance by 1°C.

stabilizer *n:* 1. a vessel in which hydrocarbon vapors are separated from liquids. 2. a fractionation system that reduces the vapor pressure so that the resulting liquid is less volatile.

stage separation *n:* an operation in which well fluids under pressure are separated into liquid and gaseous components by being passed consecutively through two or more separators. The operating pressure of each succeeding separator is lower than the one preceding it. Stage separation is an efficient process in that a high percentage of the light ends of the fluid are conserved.

still *n:* any vessel in which hydrocarbon distillation is effected.

straight-chain compound *n:* a compound in which the atoms are aligned in a straight chain.

straightening vanes *n pl:* bundles of small-diameter tubing tack-welded together in a concentric pattern and placed in the upstream section of an orifice meter run for the purpose of reducing considerably the amount of straight pipe required upstream of the orifice. They eliminate swirls and crosscurrents set up by the pipe fittings and valves preceding the meter tube.

stuffing box *n:* a device that prevents leakage along a piston rod, propeller shaft, or other moving part that passes through a hole in a cylinder or vessel. It consists of a box or chamber made by enlarging the hole and a gland containing compressed packing. On a well being artificially lifted by means of a sucker rod pump, the polished rod operates through a stuffing box, preventing escape of oil and diverting it into a side outlet to which is connected the flow line leading to the oil and gas separator or to the field storage tank. For a bottomhole pressure test, the wireline goes through a stuffing box and lubricator, allowing the gauge to be raised and lowered against the well pressure. The lubricator provides a pressure-tight grease seal in the stuffing box.

supercharge *v:* to supply a charge of air to the intake of an internal-combustion engine at a pressure higher than that of the surrounding atmosphere.

supercompressibility factor *n:* a term sometimes used to mean compressibility factor, but more often as the compressibility factor that is appropriate for high-pressure calculations. See *compressibility factor.*

sweet *adj:* having an absence or near-absence of sulfur compounds, as defined by a given specification standard.

T

tensile strength *n:* the greatest longitudinal stress that a metal can bear without tearing apart. Tensile strength of a metal is greater than yield strength.

thermocouple *n:* a device consisting of two dissimilar metals bonded together, with electrical connections to each. When the device is exposed to heat, an electrical current is generated, the magnitude of which varies with the temperature. It is used to measure temperatures higher than those that can be measured by an ordinary thermometer, such as those in an engine exhaust.

thermodynamic equilibrium *n:* property of a system that is in mechanical, chemical, and thermal equilibrium.

thermosiphon baffle *n:* a device for assisting in controlling the direction of thermal currents in an indirect heater.

throttling valve *n:* a valve used to regulate the rate of flow, usually only partially open or closed.

tight well *n:* a well that produces from a reservoir of low permeability and thus stabilizes slowly. Such wells show slow buildup of bottomhole pressure on a buildup test, and the flowing pressure does not stabilize quickly at a constant value on a flow test.

trap *n:* a body of permeable oil-bearing rock surrounded or overlain by an impermeable barrier that prevents oil from escaping. The types of traps are structural, stratigraphic, hydrodynamic, or a combination of these.

tubing *n:* small-diameter pipe that is run into a well to serve as a conduit for the passage of oil and gas to the surface.

turbine meter *n:* a velocity-measuring device for fluids in which the flow is parallel to the rotor axis and the speed of rotation is proportional to the rate of flow. The volume of gas in gas measurement is determined by counting the revolutions of the rotor. In liquid turbine meter measurement, the meter and electronic instrumentation are combined to measure total flow and/or flow rate within the piping system.

turbocharge *v:* to supercharge an engine by use of a turbocharger.

turbocharger *n:* a centrifugal blower driven by exhaust gas turbines and used to supercharge an engine.

two-phase flow *n:* a flow of a vapor phase and a liquid phase in the same pipeline.

two-stroke/cycle engine *n:* an engine in which the piston moves from top dead center to bottom dead center and then back to top dead center to complete a cycle. Thus, the crankshaft must turn one revolution, or 360 degrees.

U

unconformity *n:* 1. lack of continuity in deposition between rock strata in contact with one another, corresponding to a gap in the stratigraphic record. 2. the surface of contact between rock beds in which there is a discontinuity in the ages of rocks.

V

valve *n:* a device used to control the rate of flow in a line, to open or shut off a line completely, or to serve as an automatic or semiautomatic safety device. Those with extensive usage include the gate valve, plug valve, globe valve, needle valve, check valve, and pressure relief valve.

vaporization *n:* 1. the act or process of converting a substance into the vapor phase. 2. the state of substances in the vapor phase.

vapor-liquid equilibrium ratio *n:* the partition coefficient, usually designated as K, which is equivalent to y/xm where y is the mol fraction of a given component in the vapor phase that is in equilibrium with x, the mol fraction of the same component in the liquid phase. K is a function of temperature, pressure, and composition of the particular system.

vapor pressure *n:* the pressure exerted by the vapor of a substance when the substance and its vapor are in equilibrium. Equilibrium is established when the rate of evaporation of a substance is equal to the rate of condensation of its vapor.

velocity *n:* speed; the timed rate of linear motion.

volumetric efficency *n:* 1. actual volume of fluid put out by a pump, divided by the volume displaced by a piston or pistons (or other device) in the pump. Volumetric efficiency is usually expressed as a percentage. For example, if the pump pistons displace 300 cubic inches, but the pump puts out only 291 cubic inches per stroke, then the volumetric efficiency of the pump is 97 percent. 2. for compressors, the ratio of the volume of gas actually delivered, corrected to suction temperature and pressure, to the piston displacement.

W

water drive *n:* the reservoir drive mechanism in which oil is produced by the expansion of the underlying water and rock, which forces the oil into the wellbore. In general, there are two types of water drive: bottom-water drive, in which the oil is totally underlain by water, and edgewater drive, in which only the edge of the oil is in contact with the water.

water saturation of a reservoir *n:* the percentage of total pore space occupied by water. It is normally estimated by electric log calculations, although cores specially obtained may be analyzed for this value.

wellhead *n:* the equipment installed at the surface of the wellbore. A wellhead includes such equipment as the casinghead and tubing head. *adj:* pertaining to the wellhead (e.g., wellhead pressure).

well stimulation *n:* any of several operations used to increase the production of a well, such as acidizing or fracturing.

wet gas *n:* a gas containing water, or a gas that has not been dehydrated.

wireline *n:* a small-diameter metal line used in wireline operations; also called slick line.

work *n:* the acting of a force through distance, or the overcoming of a resistance to motion. No work is done unless motion is produced. Mathematically, work is force times distance.